Volume 5 · 1985 · Number 3

Acta Biotechnologica

Journal of microbial, biochemical and bioanalogous technology

Akademie-Verlag
Berlin

ISSN 0138-4988
Acta Biotechnol., Berlin 5 (1985) 3, 223—318
EVP 30,— M

Instructions to Authors

1. Only original papers that have not been published previously will be accepted.
   Manuscripts may be submitted in English, German or Russian (in dublicate). The name of the institute (with the full address) from which the manuscript originates should be stated below the name(s) of the authoı(s). The authors are responsible for the content of their contributions.

2. Original papers should not consist of more than 20 typewritten pages (two-lined spacing), including references, tables and figures; short original communications may have a maximum of six typewritten pages.

3. Each paper should be preceded by a summary in English and also by an English translation of the title.

4. Latin names of species as well as passages to be printed in italics for greater emphasis should be denoted by a waving line. Please use only units and symbols of the SI-system.

5. Tables may be used to shorten the text or to make understanding easier. They should be numbered consecutively throughout and supplied with a brief heading. They should not appear in the text, but should be written on separate sheets.

6. The numbers and sizes of illustrations should be limited to the essential minimum, they should be numbered consecutively and be supplied on separate sheets.
   Line drawings, including graphs and diagrams, should be drawn in black ink. Half-tone illustrations should be supplied as white glossy prints. Figure legends are to be typed in sequence on a separate sheet.
   The name(s) of the author(s) should be written on the back of each sheet.

7. References listed at the end of the contribution should contain only works quoted in the text. They should be numbered in the order in which they are first mentioned in the text.
   Please give surnames and initials of all authors, the name of the journal abbreviated according to "Chemical Abstracts – List of Periodicals", volume number, year of publication, issue number or month, first page number. Books are to be cited with full title, edition, volume number, page number, place of publication, publisher and year of publication.

8. Notes to the text may be given as footnotes on the same page.

9. 50 printed copies are free of charge.
   Additional reprints may be ordered (at full cost).

10. The author will receive two galley proofs for correction. They are to be returned to the managing editor (DDR-7050 Leipzig, Permoserstr. 15, Dr. Dimter) as soon as possible.

Acta Biotechnologica

Journal of microbial, biochemical and bioanalogous technology

Edited by the
Institute of Biotechnology of the Academy of Sciences of the G.D.R.; Leipzig
and by the
Institute of Technical Microbiology; Berlin
by
M. Ringpfeil, Leipzig and G. Vetterlein, Berlin

Volume 5

1985

Number 3

AKADEMIE-VERLAG · BERLIN

"Acta Biotechnologica" publishes original papers, short communications, reports and reviews from the whole field of biotechnology.
The journal is to promote the establishment of biotechnology as a new and integrated scientific field.
The field of biotechnology covers microbial technology, biochemical technology and the technology of synthesizing and applying bioanalogous reaction systems.
The technological character of the journal is guaranteed by the fact that papers on microbiology, biochemistry, chemistry and physics must clearly have technological relevance.

Terms of subscription for the journal "Acta Biotechnologica"

Orders can be sent

- **in the GDR:** to Postzeitungsvertrieb, or to the Akademie-Verlag Berlin, DDR-1086 Berlin, Leipziger Str. 3—4; PF-Nr. 1233;
- **in the other socialist countries:** to a book-shop for foreign languages literature or to the competent news-distributing agency;
- **in the FRG and Berlin (West):** to a book-shop or to the wholesale distributing agency Kunst und Wissen, Erich Bieber, Wilhelmstr. 4—6, D-7000 Stuttgart 1;
- **in the other Western European countries:** to Kunst und Wissen, Erich Bieber GmbH, Dufourstr. 51, CH-8008 Zürich;
- **in other countries:** to the international book- and journal-selling trade, to Buchexport, Volkseigener Außenhandelsbetrieb der DDR, DDR-7010 Leipzig, Postfach 160, or to the Akademie-Verlag Berlin, DDR-1086 Berlin, Leipziger Str. 3—4.

Acta Biotechnologica

Herausgeber: Institut für Biotechnologie der AdW,
DDR-7050 Leipzig, Permoserstr. 15
(Direktor: Prof. Dr. Manfred Ringpfeil) und

Institut für Technische Mikrobiologie
DDR-1017 Berlin; Alt-Stralau 62
(Direktor: Dipl. Ing. G. Vetterlein).

Verlag: Akademie-Verlag Berlin, DDR-1086 Berlin, Leipziger Straße 3—4;
Fernruf: 2236201 und 2236229; Telex-Nr.: 114420; Bank: Staatsbank der DDR, Berlin, Konto-Nr.: 6836-26-20712.

Redaktion: Dr. Lothar Dimter (Chefredakteur),
Käthe Geyler (Redakteur),
DDR-7050 Leipzig, Permoserstr. 15; Tel.: 2392255.

Veröffentlicht unter der Lizenznummer 1671 des Presseamtes beim Vorsitzenden des Ministerrates der Deutschen Demokratischen Republik.

Gesamtherstellung: VEB Druckhaus „Maxim Gorki", DDR-7400 Altenburg.

Erscheinungsweise: Die Zeitschrift „Acta Biotechnologica" erscheint jährlich in einem Band mit 4 Heften. Bezugspreis eines Bandes 120,— DM zuzüglich Versandspesen; Preis je Heft 30,— DM. Der gültige Jahresbezugspreis für die DDR ist der Postzeitungsliste zu entnehmen.
Bestellnummer dieses Heftes: 1094/5/3.

AN (EDV) 42133

03000

Acta Biotechnol. 5 (1985) 3, 225—234

Thermophilic Anaerobic Digestion for Sewage Sludge Stabilization

HAMER, G., BRYERS, J. D. and BERGER, J.

Swiss Federal Institute for Water Resources and
Water Pollution Control (EAWAG)
Swiss Federal Institutes of Technology
CH-8600 Dübendorf, Switzerland

Summary

The microbially mediated biochemical reactions that occur during anaerobic digestion processes for methane production from soluble carbon energy substrates are well known, but in spite of this, the interactions within the multi-species cultures responsible for the overall process require more detailed elucidation. When the process feed comprises mixed, solid, carbon energy substrates, as in the case of waste sewage sludge stabilization, many aspects of both the process biochemistry and microbiology are unresolved. This mini-review seeks to identify some of these unresolved questions, particularly with respect to operation at thermophilic temperatures.

Introduction

In almost all countries, increasingly stringent legislation concerning aqueous discharges into surface waters has resulted in increases in the volumes of sewage and wastewater undergoing treatment and in improvements in the effectiveness of treatment processes with respect to their pollutant removal performance. Conventional sewage and wastewater treatment processes comprise a series of distinct stages. In the preliminary and primary stages, coarse and fine solids are physically and mechanically separated from the influent. In the secondary stage, soluble, insoluble and immiscible biodegradable matter is biologically oxidized to form microbial biomass (i.e., sludge) which is subsequently concentrated by sedimentation, prior to either recycling or wastage. In the tertiary stage, residual pollutants are removed by various physical and chemical methods. During the secondary treatment stage, non-biodegradable matter is frequently entrapped, adsorbed on or absorbed in the sludge and, in cases where simultaneous phosphate precipitation is employed, secondary sludges can contain significant concentrations of inorganic phosphates.
Waste sludge is produced in the preliminary, primary and secondary process stages and as treatment process activity increases, in response to the enforcement of legislation, the quantity of sludge produced will also increase. Various methods for sludge treatment and disposal have been proposed, but no disposal method can be considered entirely satisfactory in every respect. Proposed disposal methods include:

I Spreading stabilized sludge on agricultural land;
II Mixing with solid waste and dumping in sanitary landfill sites;

III Incineration (with or without additional fuel);
IV Immobilization within solid matrices;
V Off-shore dumping

Of the various options for disposal, off-shore dumping is an obnoxious practice and a disposal method that is unavailable to land-locked regions; disposal in land-fill sites can result in groundwater pollution by the leachates produced; incineration results in air pollution and remote land and water pollution through precipitation, and deposition on agricultural land, even after stabilization, can result in the distribution of noxious chemicals and potentially pathogenic organisms both on the land and in run-off.
In Switzerland, the current policy with respect to sludge disposal is deposition on agricultural land after stabilization and storage, although immobilization within solid matrices might offer a more realistic long term solution for sludge disposal.
Both the degree and the type of stabilization treatment employed are open questions, and in this contribution thermophilic anaerobic sludge stabilization is examined.

Thermophily and Microbes

Microbes are a universally available natural resource, but the only basis on which they can harnessed for use in effective technological processes involves a comprehensive understanding of their physiology and biochemistry. Temperature is one very important factor affecting the physiology and biochemistry of microbes. Microbes can be broadly classified according to their optimum temperature ranges for growth, i.e., psychrophilic growth, where the optimum temperature is $< 10\,°C$, mesophilic growth, where the optimum temperature is between 15 °C and 40 °C and thermophilic growth, where the optimum temperature is $> 45\,°C$. Thermophilic microbes can be further sub-divided into the thermotolerant group that grow optimally within the range 40° to 50 °C and comprise predominantly facultative thermophiles; the moderately thermophilic group, that have their optimum temperature range for growth between 50 °C and 65 °C and a minimum temperature for growth of ca. 45 °C; the extremely thermophilic or caldo-active group that have growth optima $> 65\,°C$.
The taxonomic diversity of both moderately and extremely thermophilic microbes is more restricted than the taxonomic diversity exhibited by mesophilic microbes. Genuine thermophiles, as opposed to thermotolerant microbes, are optimally adapted for growth at elevated temperatures and are not struggling to survive at such temperatures [1]. However, many extreme thermophiles seem to be relatively fastidious with respect to their requirements for optimal growth [2], and hence, are probably poorly suited for application in practical thermophilic sludge digestion processes. As a result, only thermotolerant and moderately thermophilic microbes can be considered for application in sludge digestion processes.
Potential effects of temperature on microbes include those concerning:

I Endogenous metabolism and maintenance requirements;
II Specific death and lysis rate constants;
III The maximum specific growth rate constant;
IV The saturation constant of microbes for their growth limiting substrate.

In addition, all microbially mediated processes occur under heterogeneous conditions and the hetereogeneous nature of microbial systems is of obvious significance in any analysis of the effects of temperature on systems, because, even though microbially mediated processes involve a complex sequences of biochemical reactions, the rate-limiting step is frequently mass transfer.

Anaerobic Stabilization Process Feeds

The primary objectives of anaerobic sewage sludge stabilization processes are, irrespective of reactor configuration, to produce a residual sludge comprised of a minimum fraction of biodegradable, essentially putrefactive matter, and a maximum fraction of non-biodegradable, non-putrefactive matter. This sludge should be easily dewaterable, contain neither significant numbers of pathogenic organisms nor concentrations of toxic chemicals, be predominantly free from obnoxious odours and be of a perceived value for application to agricultural land. Optimum production of methane is of only secondary importance.

The process feed to anaerobic sewage sludge stabilization processes consists of a relatively dilute suspension of sludge derived from the primary and secondary stages of domestic and municipal sewage treatment processes. Its composition is highly complex and variable, as a result of both sewage composition and the type and effectiveness of the sewage treatment process producing the waste sludge. The relative proportions of sludge derived from primary and secondary treatment and sludge handling/pretreatment prior to anaerobic stabilization are also important factors with respect to composition. Although a waste sewage sludge is totally ill-defined with respect to individual compounds, any waste sewage sludge feed can, in general terms, be considered to consist of the following fractions:

I Non-biodegradable solids;
II Biodegradable, non-viable and viable microbes;
III Biodegradable solids of non-microbial origin;
IV Soluble biodegradable matter;
V Soluble non-biodegradable matter.

The soluble non-biodegradable solids will pass through the sludge stabilization process unchanged and will ultimately comprise the non-putrefactive fraction of the residual stabilized sludge. The biodegradable, non-viable and viable microbes are derived predominantly from microbial biomass comprising the waste secondary sludge, although a very small fraction can comprise potential pathogens present in the original sewage. In addition, biodegradable viable microbes will also be produced from the utilization of other biodegradable matter during the stabilization process and present an additional feed for subsequent biodegradation. Biodegradable solids of non-microbial origin comprise materials such as cellulose particles, present in the original sewage that have either been separated during primary sewage treatment or undergone partial degradation during secondary sewage treatment, and hence, comprises a significant fraction of the waste sludge. Soluble biodegradable matter present in many waste sludges will be virtually negligable, but where sludges have been subjected to either prolonged holding times or pretreatment, particularly thermophilic aerobic pretreatment, prior to anaerobic stabilization, significant concentrations of soluble partial biodegradation products can be present in the feed sludge, together with significant concentrations of facultative anaerobic bacteria. Such bacteria can also mediate biodegradation during anaerobic stabilization. Small concentrations of non-biodegradable soluble matter, either derived from non-biodegradable compounds or comprising residues from partially biodegradable compounds present in the original sewage will generally be found in association with waste sewage sludge.

Clearly, differentiation between the biodegradable versus non-biodegradable and the solid versus soluble feedstock fractions is important. However, additional differentiation between the biodegradable, microbial and non-microbial solids should be considered, since the former solids group undergo death, decay, and lysis, due to either autolytic mechanisms or exo-enzyme action, and the non-microbial solids can be predicted to decay in accordance with the shrinking-site concept proposed by MOREIRA et al. [3].

Microbial Process Steps and Sequences

In simple qualitative terms, the growth of a microbial monospecies on and product formation from a single carbon energy substrate can be represented by Fig. 1 [4] which has been modified to take account of the microbial death and lysis processes which are

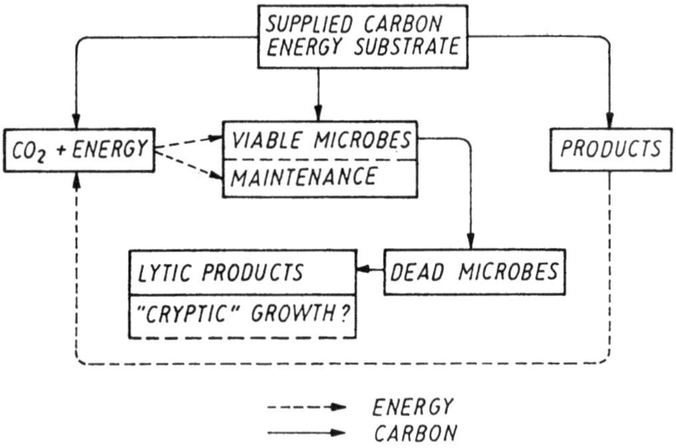

Fig. 1. Heterotrophic microbial growth — Monospecies culture.

of critical importance for effective sewage sludge stabilization. Similarly, the growth of a binary mixed microbial culture, in a completely mixed, continuous flow bioreactor on a single carbon energy substrate can be represented, diagrammatically, by Fig. 2, where death and lysis are again included and where it is also assumed that only one of the two species is able to utilize the supplied carbon energy substrate and that the

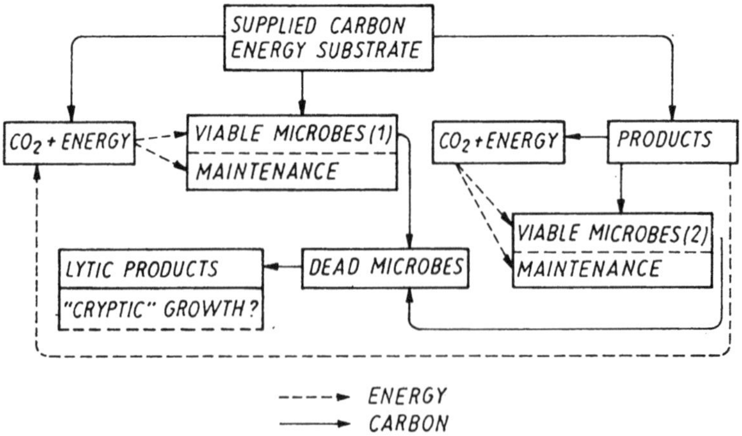

Fig. 2. Heterotrophic microbial growth — Binary culture with one species utilizing the carbon energy substrate.

second species utilizes a product produced by the first species as its carbon energy substrate. Such a binary microbial system is one of the simplest mixed culture systems possible. In practice, the supplied substrate utilizing species is usually inhibited by its products, as in the case of the aerobic methane-utilizing methanotroph/methylotroph mixed culture modelled by Wilkinson et al. [5]. However, the anaerobic utilization of complex, mixed particulate and soluble carbon energy substrates by a complex multispecies mixed culture to produce stabilized sludge, methane and carbon dioxide is infinitely more complex. The basis upon which anaerobic digestion processes function has recently been reviewed by Gujer and Zehnder [6] and by Sahm [7]. When waste sewage sludge is the process feed, the overall process consists of three major process steps, i.e., particle hydrolysis, acidogenesis and methanogenesis. However, in all three individual steps both multiple substrates and multispecies cultures are, inevitably, involved.

Particle Hydrolysis

This is the least studied and least understood step in sewage sludge stabilization. Probably the most detailed study of this step was that of Eastman and Ferguson [8] who investigated the hydrolysis of primary sludge, low in biodegradable microbial solids, to produce the soluble substrates for acidogenesis. Their results suggested that the hydrolysis of particulates to soluble substrates is the rate limiting step in the overall anaerobic digestion process, and obviously enhancement of this rate by some means, such as sludge pretreatment, would have obvious advantages with respect to reduction in the mean digestor residence times necessary to obtain an effectively stabilized residual sludge. Other major factors affecting the rate of hydrolysis of non-microbial solids are particle size and size distribution, composition of the solids and the rate of and source for appropriate enzyme production for the hydrolysis.

In the case of the hydrolysis of microbial solids, potentially different process mechanisms exist, but data concerning these mechanisms is negligeable and only crude hypotheses exist. For example, it is not even known whether death is prerequisite for lysis. Further microbial death is a concept that still needs clarification. Certainly a non-viable microbe cannot be considered to be dead; neither does posthumous oxidative endoenzyme activity constitute life.

The fundamental concept upon which waste sludge hygienization is based is one of subjecting any potentially pathogen present in the sludge to alien, essentially hostile, environmental conditions, i.e., strict anaerobiosis, elevated temperature, or a combination of the two. However, unless the mechanisms actually responsible for death are elucidated, no guarantee with respect to effective hygienization performance is possible. Clearly, extended exposure at elevated temperatures enhances the death rates of microbes, but data concerning microbial death kinetics at temperatures below 70 °C, i.e., at practical digestor operating temperatures, is sparce.

As far as non-pathogenic microbes are concerned, there are the processes that are involved in cell degradation that are the most important for the production of a stable, residual sludge, and death kinetics as such, are no longer totally dominant. The key processes in the biodegradation of microbes are death, lysis, lytic product formation, and subsequent "cryptic" growth. Microbial cell lysis can be attributed to various mechanisms, but in the context of sewage sludge digestion, autolysis, i.e., the action of endo-enzymes within the cell undergoing lysis, and lysis by exo-enzymes, produced by other microbial species present in the digestor, are probably the most important. Either mechanisms will result in the production of soluble lytic products and sub-micron size cell wall fragments. In situations where the microbes undergoing degradation are discretely dispersed, the shrinking-site hypothesis for biodegradable macrosolids hydrolysis is unlikely to apply. Where autolysis occurs, the process is within the cell and where exo-enzymes are responsible, the micro-scale of both microbial cells and cell wall fragments is such that they are already within the minimum size range for solid-liquid phase differentiation. However, for cases where the microbes under going digestion are present in compact flocs and where exo-enzyme action is the major lytic mechanism, the shrinking-site hypothesis for macro-solids degradation again becomes applicable, as does the additional question of the hydrolysis of the polysaccharide gums that mediate intermicrobial adhesion within flocs. In the case of autolysis of individual microbes present in flocs, the shrinking-site concept is again redundant, but diffusion of soluble products from the residual floc structure to the microbes that utilize the products as their carbon energy substrates cannot be ignored.

A final point with respect to the particle hydrolysis step is its probable binary nature. Most complex biodegradable solids are usually degraded first to form relatively more simple biopolymers such as lipids, carbohydrates and proteins, which are only subsequently degraded to the amino acids, simple sugars and higher molecular weight carboxylic acids that act as substrates for the acidogenesis step.

Acidogenesis

The amino acids, simple sugars and higher molecular weight carboxylic acids produced by particle hydrolysis provide the carbon energy substrates for what is described, in anaerobic digestion processes, as the acidogenic step. In addition, some of these same substrates must also be used by the microbes that produce the exo-enzymes necessary for the hydrolysis of major fractions of the particulate matter in the first process step, a process feature that is generally ignored. In addition, any microbial biomass produced during exoenzyme production, must itself be subject to lysis, but not necessarily through the action of its own exo-enzymes.

Acidogenesis comprises a series of simultaneous and sequential microbially mediated reactions in which amino acids and simple sugars are fermented, either directly or indirectly, to form acetic acid and the higher carboxylic acids are anaerobically oxidized, either directly or indirectly, to form acetic acid and/or hydrogen. Intermediate products such as propionic and butyric acids, that are also formed in the indirect routes mentioned above and are themselves converted into acetic acid and/or hydrogen during stable digestor operation, but under some operating conditions tend to accumulate in the digestor. The several substrates that are available for microbial utilization during acidogenesis are utilized by either specific species or specific groups of anaerobic bacteria. However, during operating conditions where propionate accumulates, the accumulation cannot be eliminated by introducing a feed of viable, anaerobic, propionate-utilizing bacteria. Such an operating strategy merely results in a supplementary supply of biodegradable microbial solids and does not result in any reduction of propionate levels. The several groups of bacteria responsible for acidogenesis are thought to be the microbes responsible for the largest production of new microbial biomass [6]. Both the acetate and the hydrogen produced as final products of acidogenesis are utilized in the production of methane during methanogenesis.

Methanogenesis

Methane production results from the action of two groups of methanogenic archaebacteria. One group is the acetophilic methanogens which are responsible for methane production from acetate by decarboxylation, and the other, is the hydrogenophilic methanogens, which produce methane by carbon dioxide reduction with hydrogen. Generally, the methanogenic archaebacteria have been considered to be extremely slow growing microbes, but amongst the hydrogenophilic methanogens, particularly those that grow optimally in the thermophilic temperature range, high growth rates have been observed [9]. On first sight it might seem desirable to establish conditions that allow these high growth rate organisms to predominate, but this would defeat the objectives of sludge stabilization, because acetate accumulation and ultimately process failure resulting from inhibition of various key process microbes would occur. Furthermore, ca. 70% of the methane produced is formed from acetate, and only ca. 30% from hydrogen and carbon dioxide [6].

The major disadvantage of microbial processes that are mediated by slow growing cultures is the long mean residence times needed in non-recycle, mixed continuous flow bioreactors for effective process operation. In anaerobic digestors designed for optimal methane production, either microbial biomass recycle or microbial biomass retention is practiced, with the objective of maintaining artificially high active microbial biomass concentrations, and hence, greatly enhanced process intensities [10]. Such processes usually operate with soluble feedstocks. However, for effective anaerobic sludge stabilization, neither processes operating with particulate phase recycle nor those designed on the basis of particulate phase retention are realistic. On the one hand, an unacceptable build-up of non-degradable solids will occur in the bioreactor, such that the bioreactor contents will become a thick slurry which will require high power inputs for mixing if mass transfer limitations are not to become restrictive. On the other, any treated sludge wasted from the system will contain a significant fraction of potentially biodegradable microbial solids, and hence, be a putrefactive product. The probable major advantage of slow growing microbes is the prediction that they will exhibit both significant energy requirements for maintenance and high relative death rates that will further reduce the already small microbial biomass yield coefficients exhibited by microbes growing under anaerobic conditions.

In any anaerobic mixed culture, microbially mediated process involving the sequence of process steps discussed, the original biodegradable carbon energy substrates will first be utilized to produce both microbial cells and carbonaceous products. A significant fraction of the microbial cells produced will die and lyse, thereby providing substrates for the "cryptic" growth of other microbes and the carbonaceous products will also be utilized for the growth of and product formation by other groups of microbes. Essentially, the greater the number of growth, lysis/"cryptic" growth and product formation/utilization steps involved, the lower will be the overall microbial biomass yield coefficient, and hence, the lower will be the residual microbial biomass concentration in the treated sludge and the higher will be the sum of the overall yield coefficients for stable end-product formation, i.e., those for carbon dioxide and methane. However, in no case will any residual sludge stabilized by anaerobic digestion, or any other microbial stabilization process, be completely free from potentially putrefactive matter. The putrefactive matter content is only minimized in such processes.

Enhanced Stabilization Rates

In the practical application of anaerobic digestion processes for sewage sludge stabilization using single, mixed bioreactors, the mean residence times in the bioreactor range from 15 to 40 days in the case of mesophilic (30–35 °C) operation and from 5 to 10 days in the case of thermophilic (50–60 °C) operation. In the case of similar bioreactors, longer residence times result in larger bioreactors per unit volume of waste sludge treated. However, in the case of capital cost comparisons between the bioreactors suitable for mesophilic operation and those suitable for thermophilic operation, it should be noted that different materials of construction, particularly insulation materials, invalidate conventional approaches based on scale-up rules and geometrical similarity.

An assumption that is invariably made in any discussion of anaerobic digestion processes is that the main process steps are essentially identical, irrespective of the operating temperature employed. Clearly, if this is so, the composition of residual treated sludges from mesophilic and thermophilic processes are unlikely to vary substantially as far as the relative ratio of the potentially putrefactive to the non-putrefactive fractions is concerned, provided the waste sludges fed to each process are identical. However, process operating temperature undoubtedly affects residual treated sludge structure, and hence, the ease with which the sludge can be concentrated and dewatered.

In all heterogeneous process systems, the operating temperature significantly affects the physical properties of the system. Such changes in physical properties result in modifications to mass transfer resistance which, in turn, results in switches from one controlling resistance to another in complex systems. The physical factors encountered in anaerobic digestors that are temperature sensitive include: gas solubilities, suspension viscosity and molecular diffusivities. In microbial processes, temperature also clearly exerts effects on both the biochemical and physiological mechanisms that control cell growth, product formation, cell death and cell lysis.

If it is assumed that microbial growth occurs in accordance with Monod type saturation kinetics, both the maximum specific growth rate constant, μ_m, and the saturation constant, K_s, i.e., the residual substrate concentration when $\mu = \mu_m/2$, can be expected to be affected by temperature. For chemical reactions, the effect of temperature on a reaction rate can be described by an ARRHENIUS type equation. This implies that the reaction rate is an ever-increasing function of temperature. However, it has been pointed out by CHARACKLIS and GUJER [11] that ARRHENIUS type equations can only be rigorously applied to either a single reaction or a sequence of reactions where the first reaction in the sequence is rate-limiting. Even so, ARRHENIUS Type equations are commonly used to describe the effects of temperature on microbially mediated processes, but such an approach is only strictly valid when it involves a combination of two ARRHENIUS expressions for comparison between process rates at two different temperatures.

One of the few definitive studies on the effects of temperature on microbial growth processes was that undertaken by TOPIWALA and SINCLAIR [12] with a chemostat culture of a facultative, mesophilic bacterial monospecies growing aerobically on a single carbon energy substrate in the mesophilic temperature range. Their findings were that the effect of temperature on the saturation constant, K_s, can be represented by an ARRHENIUS type expression and that the effect of temperature on the maximum specific growth rate constant, μ_m, is given by the difference between two ARRHENIUS type expressions. In addition, they also reported that the effect of temperature on the specific endogeneous rate constant is represented by an ARRHENIUS type expression. Whilst it can probably be assumed that similar temperature relationships exist for specific microbial species within their particular temperature ranges for growth, the question of evaluation of the effect of temperature on the rate of a particular microbially mediated process at a temperature in the thermophilic range with respect to a temperature in the mesophilic range, when different microbial strains are involved in each temperature range, still requires resolution [2] and the plausibility of extrapolating rate data from one temperature range to another, with disregard for the microbial stains mediating the process in the two temperature range, is questionable. However, evidence does exist that some microbes that can be isolated from thermophilic processes have very high maximum specific growth rate constants [13] and it is also proven that enzymic hydrolysis rates for solid substrates are enhanced at elevated temperatures [3].

Within any particular temperature range for growth, not all the microbes with temperature optima within that range necessarily have identical optimum temperatures. In addition, the temperature optima for maximum growth of and product formation rates by particular species are probably different. Undoubtedly, both of these factors are significant when an optimization of multi-species/multi-substrate microbial processes is sought.

As mentioned previously, Topiwala and Sinclair [12] found that the effect of temperature on the specific endogenous rate constant for a mesophilic, bacterial monospecies growing in chemostat culture is represented by an Arrhenius type expression. However, their study failed to differentiate between spontaneous death and lysis and true maintenance requirements by the growing culture so that their reported endogenous rate constant was actually a composite constant comprizing rate constants for several mechanisms with similar overall effects on the kinetic equations for growth. The generally accepted view that endogenous activity increases with increasing temperature can be explained on the basis of the effects of increasing temperature on death and lysis rather than on any effect on maintenance requirements. In continuous cultures of microbial monospecies it is possible to differentiate between true maintenance requirements and cell death and lysis, although differentiation between death and lysis is more difficult. Present opinion is that, with the exception of microbes growing at very low growth rates, maintenance requirements are constant and independent of microbial growth rates, whilst death/lysis rates are growth rate dependent as demonstrated by Drozd et al. [14].

In thermophilic sludge digestion processes, the question of lysis applies to both the thermophilic process microbes and the mesophilic microbes present in the waste sludge undergoing treatment. As far as the latter are concerned, the relatively mild temperature shock induced by the change from an ambient temperature of 15—20 °C to a digestor operating temperature of 50—60 °C can be expected to promote susceptibility to lytic enzyme activity and autolytic mechanisms [15]. In addition, the higher the digestor operating temperature, the higher will be the death rate of mesophilic microbes when fed into the digestor.

Clearly, temperature is a most important factor in enhancing the potential performance of waste sludge stabilization digestor. The critical factor with respect to the thermophilic operation is the available spectrum of process microbes suited to optimum performance within the thermophilic temperature range. The lack of adequate direct heat production during anaerobic digestion reactions inevitably means that methane produced in the digestion process must be used to heat the process. This is not necessarily a disadvantage in processes where effective waste sludge stabilization and hygienization, rather than maximum biogas production, is the primary process objective.

Acknowledgement

Financial support for this work was provided by Swiss National Programme 7B.

Received July 13, 1984

References

[1] Brock, T. D.: Proc. 19th Symp. Soc. Gen. Microbiol., London. 1969, p. 15.
[2] Sonnleitner, B., Fiechter, A.: Trends Biotechnol. **1** (1983), 74.
[3] Moreira, A. R., Phillips, J. A., Humphrey, A. E.: Biotechnol. Bioeng. **23** (1981), 1325.
[4] Stouthamer, A. H.: Vierteljahresschrift der Naturforschenden Gesellschaft in Zürich **125** (1980), 43.
[5] Wilkinson, T. G., Topiwala, H. H., Hamer, G.: Biotechnol. Bioeng. **16** (1974), 41.
[6] Gujer, W., Zehnder, A. J. B.: Water Sci. Technol. **15** (1983), 127.
[7] Sahm, H.: Adv. Biochem. Eng. **29** (1984), 83.

[8] EASTMAN, J. A., FERGUSON, J. F.: J. Wat. Poll. Control Fed. **53** (1981), 352.
[9] ZEHNDER, A. J. B., INGVORSEN, K., MARTI, T. — In: Anaerobic Digestion. Eds.: HUGHES et al. Amsterdam: Elsevier, NL. 1982, p. 45.
[10] STUCKEY, D. C. — In: Management of Industrial Wastewater in Developing Nations. Eds.: STUCKEY and HAMZA. Oxford: Pergamon, GB. 1982, p. 200.
[11] CHARACKLIS, W. G., GUJER, W.: Prog. Wat. Technol. **11** (1979) 1, 111.
[12] TOPIWALA, H. H., SINCLAIR, C. G.: Biotechnol. Bioeng. **13** (1971), 795.
[13] SONNLEITNER, B., FIECHTER, A.: Eur. J. Appl. Microbiol. Biotechnol. **18** (1983), 174.
[14] DROZD, J. W., LINTON, J., DOWNS, J., STEPHENSON, R. D.: FEMS Microbiol. Lett. **4** (1978), 311.
[15] CARENBERG, C.-O., HEDEN, C.-G.: Biotechnol. Bioeng. **12** (1970), 167.

Book Review

Wulf CRUEGER, Anneliese CRUEGER

Biotechnologie — Lehrbuch der angewandten Mikrobiologie

2. Auflage/Studienausgabe.

München, Wien: R. Oldenbourg Verlag 1984. 320 S., 240 Abb., 124 Tab., 58 DM

Die Mikrobiologie hat in den letzten Jahren in der Anwendung immer größere Bedeutung gewonnen. Mikrobiologische Methoden sind wichtig bei der Herstellung von Pharmazeutika, in der Biotechnologie, der Lebensmitteltechnik und auch der Tier- und Pflanzenzüchtung. Das Lehrbuch beschreibt die Eigenschaften der Mikroorganismen und die für die Praxis bedeutenden mikrobiologischen Methoden und Techniken. Es richtet sich bevorzugt an Dozenten und fortgeschrittene Studenten der Mikrobiologie — und es wird ebenso dem Praktiker in der pharmazeutischen, biologischen und biotechnologischen Forschung als informierendes Nachschlagewerk dienen.
Aus dem Inhalt —
Einleitung; Screening auf neue Metabolite; Stammentwicklung; Substrate für die industrielle Fermentation; Fermentationstechnik; Aufarbeitung; Organische Rohstoffe durch Gärungen; Organische Säuren; Aminosäuren; Nucleoside, Nucleotide und verwandte Verbindungen; Enzyme; Vitamine; Antibiotika; Mutterkornalkaloide; Mikrobielle Transformationen; Single Cell Protein (SCP); Neuere Verfahren in der Abwasserbehandlung; Leaching; Extracelluläre Polysaccharide; Weitere Fermentationsverfahren und Ausblick; Ergänzung zur 2. Auflage.

Acta Biotechnol. 5 (1985) 3, 235–243

Experimentelle Untersuchung, Modellierung und optimale Berechnung eines mehrstufigen Säulenfermentors

I. Theoretische Analyse und Synthese des mathematischen Modells

Kafarov, V. V.[1], Vinarov, A. J.[2], Gordeev, L. S.[1] und Semenova, E. A.[2]

[1] Moskauer chemisch-technologisches D. I. Mendelejew-Institut
125820 Moskau, Miuskaja pl. d. 9, UdSSR
[2] Allunionsinstitut für Biosynthese von Eiweißstoffen,
109004 Moskau, B. Kommunistitscheskaja 27, UdSSR

Summary

A model of the process of aerobic fermentation in a tower fermenter devided in sections was established and examined by experiments.
In this part data for the concentration of biomass, substrate and oxygen in the sections of the bioreactor at countercurrent moving of the phases were characterized.

Auf der Grundlage einer Systemanalyse wurde ein Modell des Prozesses der aeroben Fermentation für einen mehrstufigen Säulenfermentor erarbeitet und experimentell überprüft. Es wurden Daten zur Kinetik und Stöchiometrie des Kultivierungsprozesses der Mikroorganismen, zum Einfluß der Gasgeschwindigkeit auf den Sauerstoffübergang sowie zur Struktur der Ströme in Labor- und Versuchsapparaten ermittelt. Ein Algorithmus zur Bewertung der für die Funktion des mehrstufigen Säulenfermentors ausschlaggebenden Parameter wurde aufgestellt.
Dieser Algorithmus wird verwendet, um Probleme der optimalen Auslegung zu klären.
Im ersten Teil des Artikels werden die Hauptblöcke des mathematischen Modells des Fermentors insgesamt beschrieben. Es wird ein Gleichungssystem für das mathematische Modell aufgestellt, das die Verteilung der Biomasse-, Substrat- und Sauerstoff-Konzentration über die Sektionen des Apparates bei einer Gegenstrombewegung der Phasen charakterisiert.
Infolge der Dringlichkeit des Aufbaus einer Eiweißgroßproduktion mit einer Jahreskapazität von 100000 t und darüber gewinnen Probleme der Herstellung von Hochleistungsfermentoren zunehmend an Aktualität. Zum gegenwärtigen Zeitpunkt gibt es eine Vielzahl von Fermentoren, die hinsichtlich ihres Arbeitsprinzips und ihrer konstruktiven Merkmale Unterschiede aufweisen. Von besonderem Interesse sind dabei die Säulenfermentoren, die trotz einer verhältnismäßig einfachen Konstruktion zuverlässige, hochproduktive Apparate darstellen.
Für eine optimale Auslegung derartiger Bioreaktoren ist es erforderlich, sichere Methoden zur Modellierung und Berechnung des Fermentationsprozesses, auch unter Berücksichtigung der Stoffübergangswerte und hydrodynamischen Kennziffern zu erarbeiten. Das aufzustellende mathematische Modell des Fermentors sollte die komplizierte innere Wechselbeziehung zwischen physikalisch-chemischen Prozessen einerseits und biochemischen Prozessen andererseits widerspiegeln, was nur durch die gewählte

Methodik einer Systemanalyse [1, 2] möglich ist, die häufig zum Studium technologischer Prozesse eingesetzt wird.
Es sind eine Reihe von Arbeiten bekannt, die sich mit der Untersuchung und mathematischen Modellierung einzelner Elemente des in unterschiedlichsten Bioreaktoren, darunter auch in Säulenfermentoren [3—10], ablaufenden Fermentationsprozesses beschäftigen. Die experimentellen Untersuchungen werden aber in diesen Arbeiten vorzugsweise an Labormodellen mit einem Arbeitsvolumen von nur einigen Litern durchgeführt, wodurch kein Urteil darüber gefällt werden kann, ob diese mathematischen Modelle zur Berechnung von industriellen Fermentoren praktikabel sind.
Die Hauptaufgabe der von uns durchgeführten Untersuchungen ist die Erarbeitung einer Gesamtstrategie zum Aufbau eines mathematischen Modells für einen biochemischen Reaktor sowie die Verwendung dieses mathematischen Modells zur Analyse und Optimierung der Arbeitsparameter von mehrstufigen Säulenfermentoren mit unterschiedlichem Fassungsvermögen.
Wenn man die Prozesse, die in einem einzelnen Bioreaktor ablaufen, untersucht, kann man die Gesamtheit der dabei auftretenden Erscheinungen in einige hierarchische Stufen unterteilen, die die Mikro- und Makrokinetik charakterisieren. Zu den Faktoren, die für die Mikrokinetik ausschlaggebend sind, gehört ein Komplex an physikalischen, chemischen und biochemischen Erscheinungen, die auf das Niveau der Mikroorganismenzellen und -populationen konzentriert sind. Zu den Faktoren, die die Makrokinetik kennzeichnen, zählen hydrodynamische Effekte, Wärme- und Diffusionseffekte mit den charakteristischen Merkmalen des großindustriellen Maßstabes, deren Struktur sich vor allem aus den konstruktiven Besonderheiten des Apparates, aus der Art und dem Verfahren des Energieeintrages, aus der Wärmeableitung u. ä. ergibt. Einer Systemanalyse folgend, lassen sich nachfolgende hierarchische Stufen, die das genannte System charakterisieren, aufstellen. Die erste Hierarchiestufe erfaßt die Bedingungen für die Funktion der Einzelzellen bei deren Entwicklung im Medium mit örtlich verteilten Parametern, die die Menge an Nährstoffen, Sauerstoff u. a. definieren, die unmittelbar an der Zellmembran lokalisiert ist. Die zweite Stufe beschreibt einen Komplex an Effekten in der Zellpopulation unter Berücksichtigung von Aggregaten und Globuli, in denen die Entwicklungsbedingungen für die Einzelzellen unterschiedlich sind.
Die dritte Stufe charakterisiert den Stoff-, Energie- und Impulstransport über die Phasengrenze in ein heterogenes Medium. Die im Ergebnis von Übertragungseffekten auftretenden Deformationen an der Phasengrenzfläche resultieren aus Prozessen der Dispergierung von Substrattropfen und Gasblasen und einer Veränderung der Phasengrenzfläche. Die Erscheinungen der vierten Stufe determinieren den hydrodynamischen Zustand und die Prozesse des Stoff- und Energie-Übergangs in einem Volumenelement des Apparates.
Die fünfte hierarchische Strukturstufe stellt einen Komplex an Erscheinungen dar, die den Zustand der Hydrodynamik und des Stoffübergangs im Apparat unter Berücksichtigung seiner konstruktiven Besonderheiten, des Typs der Misch- und Belüftungsvorrichtungen, des Verfahrens, des Energieeintrages u. ä. definieren.
Die betrachteten Stufen in ihrer Komplexität gestatten es, ein mathematisches Modell des Fermentors insgesamt aufzustellen, das eine konkrete Variante der Organisation der Stoff- und Energieströme in dem System wiedergibt. Abb. 1 zeigt die prinzipielle Grobstruktur des Modells des biochemischen Reaktors einschließlich der oben beschriebenen Hauptblöcke des Modells.
Entsprechend dieser Struktur schließt die Modellierung der Kinetik des Kultivierungsprozesses der Mikroorganismen in einem biochemischen Reaktor die Erarbeitung eines kinetischen Wachstumsmodells der Zellpopulation ein, das die Wachstumsgeschwindigkeit und die Absterberate der Einzelzellen berücksichtigt. Diese Stufe ist mikrokineti-

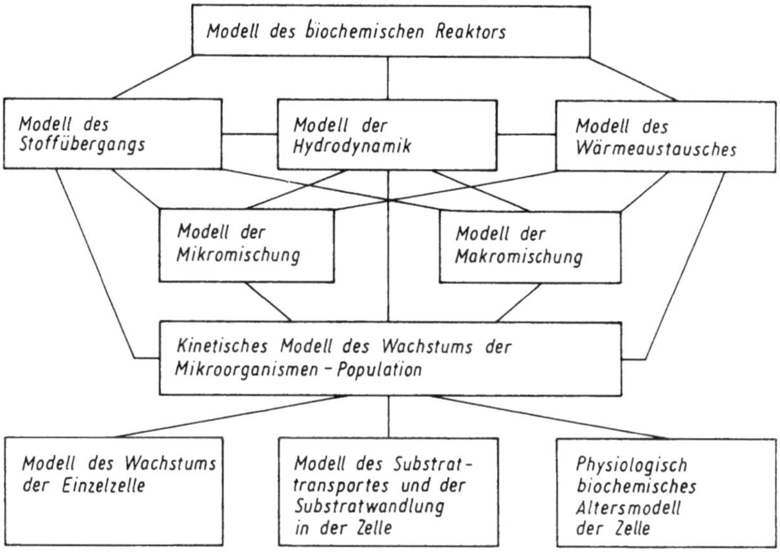

Abb. 1. Grobstruktur des Modells eines Bioreaktors

scher Natur, die Stufen darüber beziehen sich auf die Makrokinetik und sind im Schema durch folgende Blöcke dargestellt:
durch das Modell der Mikromischung und das Modell der Makromischung, die ihrerseits das Modell der Hydrodynamik im Fermentor bilden, wobei letzteres die Effekte zusammenfassend berücksichtigt. Die konkrete Art der Gleichungen der Modelle des Stoff- und Wärmeübergangs werden durch die Bedingungen der Mikro- und Makromischung determiniert, da diese Prozesse den Stoff- und Energietransport unmittelbar beeinflussen. Das Gesamtmodell des biochemischen Reaktors, das auf der Grundlage der Modelle der Hydrodynamik und des Stoff- und Wärmeübergangs unter Berücksichtigung des kinetischen Wachstumsmodells der Mikroorganismen-Population aufgestellt wird, kann zur Optimierung der technologischen und konstruktiven Parameter von Bioreaktoren sowie für deren optimale Projektierung angewendet werden.

Kinetisches Modell

Als Hauptfaktoren sollte das kinetische Modell den Einfluß der Substrat-, Gelöstsauerstoff- und Biomasse-Konzentration auf die Wachstumsgeschwindigkeit und Stöchiometrie des Kultivierungsprozesses der Mikroorganismen berücksichtigen. Für das Gleichungssystem des Modells (für das ideal gemischte Volumenelement V_i) gilt:

$$\begin{aligned}
\frac{dX}{dt} &= \mu_m \frac{S}{K_1 + S} \cdot \frac{K_2}{K_2 + S_0 - S} \cdot X - D \cdot X \\
\frac{dS}{dt} &= D(S_0 - S) - \alpha^S \mu_m \frac{S}{K_1 + S} \cdot \frac{K_2}{K_2 + S_0 - S} \cdot X \\
\frac{dC_L}{dt} &= k_L a(C^* - C_L) - \alpha^{O_2} \mu_m \frac{S}{K_1 + S} \cdot \frac{K_2}{K_2 + S_0 - S} \cdot X - D \cdot C_L \\
D &= \frac{V}{V_i}
\end{aligned} \tag{1}$$

Dabei bedeuten:

X — Mikroorganismenkonzentration $\left[\frac{g}{l}\right]$

S — Substratkonzentration $\left[\frac{g}{l}\right]$

C_L — Gelöstsauerstoffkonzentration $\left[\frac{g}{l}\right]$

μ_m — maximale spezifische Wachstumsgeschwindigkeit $[h^{-1}]$

D — Durchflußrate $[h^{-1}]$

$k_L a$ — volumenbezogener Stoffübergangskoeffizient $[h^{-1}]$

C^* — Sauerstoffsättigungskonzentration $\left[\frac{g}{l}\right]$

α^S — Substratverbrauchskoeffizient $\left[\frac{g}{g}\right]$

α^{O_2} — Sauerstoffverbrauchskoeffizient $\left[\frac{g}{g}\right]$

K_1, K_2 — kinetische Konstanten $\left[\frac{g}{l}\right]$

Nachfolgende Gleichung berücksichtigt den Einfluß der Gelöstsauerstoff-Konzentration auf den Wert von μ:

$$\mu = \mu_m \frac{\bar{C}_L}{K_3 + \bar{C}_L + \frac{\bar{C}_L^2}{K_4}} \tag{2}$$

K_3, K_4 — kinetische Konstanten

μ — spezifische Wachstumsgeschwindigkeit $[h^{-1}]$

Die Gleichungen (3a) und (3b) zeigen die Abhängigkeit der stöchiometrischen Koeffizienten α^S, α^{O_2} von der Wachstumsgeschwindigkeit der Mikroorganismen:

$$\alpha^S = a_1 + \frac{b_1}{\mu} \tag{3a}$$

$$\alpha^{O_2} = a_2 + \frac{b_2}{\mu} \tag{3b}$$

Dabei sind:

a_1, a_2 — stöchiometrische Konstanten $\left[\frac{g}{g}\right]$

b_1, b_2 — stöchiometrische Konstanten $\left[\frac{g}{g \cdot h}\right]$

Modell der Makromischung

Bei der Erarbeitung eines mathematischen Modells, das die Effekte der Makromischung berücksichtigt, ist es erforderlich, auf eine konkrete apparative Konfiguration des Fermentationsprozesses in Säulenfermentoren mit einem Durchmesser von 100 und 600 mm und einer Höhe von 300 und 7000 mm zu orientieren [11].

Abb. 2 zeigt eine Prinzipskizze des mehrstufigen Säulenfermentors. Der Apparat ist mittels dynamischer Böden in Kammern unterteilt, in denen eine Schaumbekämpfung des Mediums erfolgt. In den Kammern des Apparates sind Füllkörper aus einem inerten Material angebracht, die eine effektive Dispergierung der Luft und eine günstige Turbulenz des Mediums bewirken.

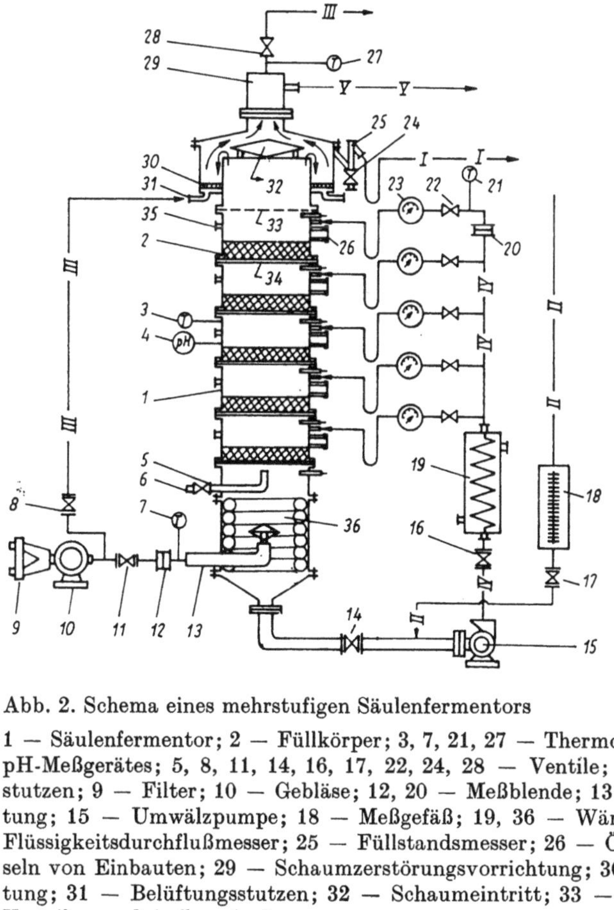

Abb. 2. Schema eines mehrstufigen Säulenfermentors

1 — Säulenfermentor; 2 — Füllkörper; 3, 7, 21, 27 — Thermometer; 4 — Geber des pH-Meßgerätes; 5, 8, 11, 14, 16, 17, 22, 24, 28 — Ventile; 6, 35 — Probenahmestutzen; 9 — Filter; 10 — Gebläse; 12, 20 — Meßblende; 13 — Belüftungsvorrichtung; 15 — Umwälzpumpe; 18 — Meßgefäß; 19, 36 — Wärmeaustauscher; 23 — Flüssigkeitsdurchflußmesser; 25 — Füllstandsmesser; 26 — Öffnung zum Auswechseln von Einbauten; 29 — Schaumzerstörungsvorrichtung; 30 — Dispergiereinrichtung; 31 — Belüftungsstutzen; 32 — Schaumeintritt; 33 — Rückhaltesieb; 34 — Verteiler- und Auflagerboden

I — Kulturflüssigkeit
II — Nährsalzlösung
III — Luft
IV — rezirkulierende Kulturflüssigkeit
V — Biomasse

Die Luft wird in die untere Kammer eingeleitet und gleichmäßig über den Querschnitt der Säule verteilt. Durch Einsatz einer Pumpe sorgt der zirkulierende Strom für die notwendige Verteilung der Komponenten auf die einzelnen Kammern des Apparates, und es bildet sich eine hydrodynamische Struktur der Gas- und Flüssigkeitsphase heraus.

Effekte der Entmischung des Fermentationsmediums können infolge der Turbulenz des Mediums und einer effektiven Dispergierung in der vorgestellten Konstruktion des Bioreaktors vernachlässigt werden, und es kann von den Bedingungen einer maximal möglichen Durchmischung in jeder Kammer der Säule ausgegangen werden. Zur Beschreibung der Struktur der Flüssigkeitsströme durch die mehrstufige Säule verwendete man entsprechend dem Schema (Abb. 3) ein Stufenmodell mit Rückführung und ein Stufenmodell ohne Rückführung für die Gasphase:

$$V_i\left(\frac{\mathrm{d}X_i}{\mathrm{d}t}\right) = F(1+\beta)\cdot X_{i-1} + F\cdot\beta\cdot X_{i+1} - F\cdot\beta\cdot X_i - F(1+\beta)\cdot X_i$$

$$V_i\left(\frac{\mathrm{d}S_i}{\mathrm{d}t}\right) = F(1+\beta)\cdot S_{i-1} + F\cdot\beta\cdot S_{i+1} - F\cdot\beta\cdot S_i - F(1+\beta)\cdot S_i$$

$$V_i\left(\frac{\mathrm{d}C_{L_i}}{\mathrm{d}t}\right) = F(1+\beta)\cdot C_{L_{i-1}} + F\cdot\beta\cdot C_{L_{i+1}} - F\cdot\beta\cdot C_{L_i} - F(1+\beta)\cdot C_{L_i} + V_i\cdot k_La\cdot\left[\frac{y^{O_2}\cdot P}{m_{O_2}} - C_{L_i}\right] \tag{4}$$

$$\frac{\mathrm{d}y^{O_2}}{\mathrm{d}z} - \frac{H}{W}\,k_La^{O_2}\left(\frac{y^{O_2}\cdot P}{m_{O_2}} - C_L\right) = 0$$

$$\frac{\mathrm{d}y^{CO_2}}{\mathrm{d}z} + \frac{H}{W}\,k_La^{CO_2}\left(C_L{}^{CO_2} - \frac{y^{CO_2}\cdot P}{m_{CO_2}}\right) = 0$$

F — Strom durch den Bioreaktor
t — Zeit
β — Anteil des Rückflußstromes
y — Sauerstoffgehalt in der Gasphase
m — Konstante des Phasengleichgewichtes
H — Höhe der Säule [m]
W — Gasgeschwindigkeit $\left[\frac{\mathrm{m}}{\mathrm{s}}\right]$

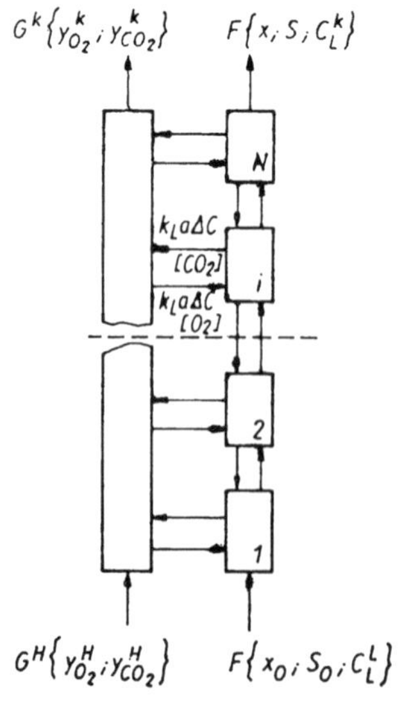

Abb. 3. Strukturschema eines Stufenmodells mit Rückführung

Als V_i wurde das Arbeitsvolumen jeder Kammer der Säule (als geschlossenes Ganzes) angenommen. Die Teilgröße des Rückflußstromes β wurde als Verhältnis des Rückflußstromes zwischen den Kammern des Säulenfermentors zum Zulaufstrom in den Apparat definiert.

Modell des Stoffüberganges

Der Einfluß der Geschwindigkeit des zur Belüftung eingesetzten Gases und des zirkulierenden Stromes auf den Sauerstoff-Übergang wurde durch die Gleichung (5) erfaßt:

$$\mathrm{Nu} = C_1 \cdot \mathrm{Re}^m \cdot \mathrm{Pr}^n (1 + f) \tag{5}$$

f — Faktor des hydrodynamischen Zustandes des Gas-Flüssigkeits-Systems [1], das vom Wechselverhältnis der Gas- und Flüssigkeitsbelastungen im Säulenfermentor abhängig ist.

$$\mathrm{Nu} = \frac{k_L a \cdot l^2}{D_F}; \qquad \mathrm{Re} = \frac{W \cdot l \cdot \varrho_F}{\mu_F}$$

$$\mathrm{Pr} = \frac{\mu_F}{\varrho_F \cdot D_F}; \qquad l = \sqrt{\frac{\sigma}{\varrho_F \cdot g}}$$

Nu — Nusselt-Zahl

Re — Reynolds-Zahl

Pr — Prandtl-Zahl

σ — Oberflächenspannung [N/m]

ϱ_F — Dichte $\left[\frac{\mathrm{kg}}{\mathrm{m}^3}\right]$

μ_F — dynamische Viskosität $\left[\mathrm{N} \cdot \frac{\mathrm{s}}{\mathrm{m}^2}\right]$

g — Beschleunigung des freien Falles $\left[\frac{\mathrm{m}}{\mathrm{s}^2}\right]$

Gleichung (6) gibt den Einfluß der hydrodynamischen Flüssigkeitssäule und des Überdrucks auf die Gelöstsauerstoff-Sättigungskonzentration in jeder Kammer in einem rückvermischungsfreiem Regime für die Gasphase an.

$$C^*_{O_2 i} = \frac{y^{O_2}}{m_{O_2}} \left[\frac{\eta \cdot (i/N)}{2{,}3 \lg \dfrac{1}{1 - \eta \cdot (i/N)}} \right] \left[P_N + \Delta P_N \frac{H}{N} (N - i) \right] \tag{6}$$

In dieser Gleichung bedeuten:

η — Ausnutzungsgrad des Sauerstoffs
y^{O_2} — Sauerstoffanteil im Eingangsgas
m_{O_2} — Konstante des Phasengleichgewichtes
N — Anzahl der Kammern
P_N — Druck am Ausgang
H — Arbeitshöhe der Säule

Unter Berücksichtigung von (1—6) lautet dann das Gleichungssystem des Modells des Säulenfermentors für die aktuelle Biomasse-, Substrat- und Sauerstoff-Konzentration der Kammer i):

$$X_{i-1} = X_i + \frac{\beta}{1+\beta} \cdot X_i - \frac{\beta}{1+\beta} \cdot X_{i+1} - \frac{\tau}{1+\beta} \left[\mu_m \frac{S_i}{K_1 + S_i} \cdot \frac{K_2}{K_2 + S_0 - S_i} \cdot X_i \right]$$

$$S_{i-1} = S_i + \frac{\beta}{1+\beta} \cdot S_i - \frac{\beta}{1+\beta} \cdot S_{i+1} + \frac{\tau}{1+\beta} \left[a_1 \mu_m \frac{S_i}{K_1 + S_i} \cdot \frac{K_2}{K_2 + S_0 - S_i} \cdot X_i + b_1 \cdot X_i \right] \tag{7}$$

$$C_{L_{i-1}} = C_{L_i} + \frac{\beta}{1+\beta} \cdot C_{L_i} - \frac{\beta}{1+\beta} C_{L_{i+1}} - \frac{\tau}{1+\beta} \cdot k_L a \left\{ \frac{y^{O_2}}{m_{O_2}} \cdot \left[\frac{\eta(i/N)}{2{,}3\, lg \frac{1}{1 - \eta(i/N)}} \right] \cdot \left[P_N + \Delta P_N \frac{H}{N} (N - i) \right] - C_{L_i} \right\} + \frac{\tau}{1+\beta} \cdot \left[a_2 \mu_m \frac{S_i}{K_1 + S_i} \cdot \frac{K_2}{K_2 + S_0 - S_i} \cdot X_i + b_2 X_i \right]$$

Dabei bedeutet τ — Verweilzeit [h].
Der Ausstoß des Bioreaktors wird hierbei an Hand der Formel (8)

$$A = \frac{X_N}{\tau} \cdot \sum_{i=1}^{N} V_1 \tag{8}$$

unter Berücksichtigung der Limitierung durch die Restsubstrat-Konzentration

$$S_N \leqq S$$

ermittelt. Das Arbeitsregime des Apparates soll die erforderlichen Voraussetzungen seitens der Hydrodynamik und des Stoffübergangs gewährleisten, um ausgehend von der Beziehung (9) den Sauerstofftransport in die Kulturflüssigkeit zu ermöglichen:

$$k_L a \left[\frac{y^{O_2} \cdot P}{m_{O_2}} - C_i \right] \geqq (a_2 \cdot \mu \cdot X_i + b_2 X_i) \tag{9}$$

Dabei wird die minimale Arbeitskonzentration für Gelöstsauerstoff entsprechend der Bedingung $C_1 \geqq C_{\text{krit.}}$ ausgewählt.
In den nachfolgenden Teilen des Artikels werden ein Algorithmus zur Berechnung des Säulenfermentors und die Hauptergebnisse einer Rechneranalyse des Modells im Vergleich zu den experimentellen Daten behandelt. Desweiteren wird die Problematik der Auswahl optimaler Parameter bei der Projektierung industrieller Säulenfermentoren auf der Basis des technisch-ökonomischen Kriteriums der Optimalität untersucht.

$$\Phi = \sum l^j \cdot d^j + \frac{N_g + N_m}{0{,}785 \cdot X_{\frac{N}{\tau}} \cdot D^2 \cdot H \cdot (1 - \bar{\varphi})} \cdot e + \frac{E \cdot K}{0{,}785 \cdot X_{\frac{N}{\tau}} \cdot D^2 \cdot H \cdot (1 - \bar{\varphi})} \rightarrow \min$$

Dabei sind:

b^j — Preis der Zulaufkomponente j
α^j — Spezifischer Verbrauch der Zuflußkomponente j pro Biomasseeinheit
N_g, N_m — Energieaufwand für Belüftung und Mischen
E — Rückflußkoeffizient der Investitionen
K — Investitionsaufwand
D, H — Durchmesser und Höhe des Säulenfermentors
$\bar{\varphi}$ — mittlerer Gasgehalt im Apparat
e — Elektroenergiepreis

Übersetzung aus dem Russischen: D. Fiedler

Eingegangen: 27. 10. 1983

Literatur

[1] KAFAROV, V. V., DOROCHOV, I. N.: Systemanalyse von Prozessen der chemischen Technologie. Moskau, 1976. S. 489.
[2] KAFAROV, V. V., VINAROV, A. J., GORDEEV, J. S.: Modellierung von biochemischen Reaktoren. Moskau, 1979. S. 343.
[3] MACLEAN, G., ERICKSON, L., USU, K., FAN, L.: Biotechnol. Bioeng. **19** (1977), 493.
[4] SHIOJA, S., DANG, N., DUNN, I.: Chem. Eng. Sci. **33** (1978), 1025.
[5] VINAROV, A. J., KAFAROV, V. V., ŠERSTOBITOV, V. V., GORDEEV, L. S.: Säulenfermentoren für mikrobiologische Prozesse. Moskau, 1976. S. 48.
[6] BUCHHOLZ, H., LUTTMANN, R., ZAKRZEWSKI, W., SCHÜGER, K.: Chem. Eng. Sci. **35** (1980), 111.
[7] FUKUDA, H., SHIOTANI, T., OKADA, W., MORIKAWA, H.: J. Ferment. Technol. **56** (1978) 6, 619.
[8] DANG, N., DUNN, I.: 4th International Congress on Chemical Engineering in Scandinavia. 1977. 15.
[9] SERIEYS, M., GOMA. G., DURAND, G.: Biotechnol. Bioeng. **20** (1978), 1393.
[10] VINAROV, A. J., SEMENOVA, E. A., GORDEEV, L. S., KAFAROV, V. V.: 2. Allunionstagung der AdW der UdSSR „Theorie und Praxis der kontinuierlichen Kultivierung von Mikroorganismen". Moskau, 1978. S. 118.
[11] VINAROV, A. J.; ŠERSTOBITOV, V. V.: Chimitscheskaja promyschlennost (1981) 9, 39.

Acta Biotechnol. 5 (1985) 3, 244

Book Review

A. Fiechter

Chromatography

Berlin: Akademie-Verlag, 1983. 138 S., 41 Abb., 32 Tab., 68 M

Der Band ist eine Sonderausgabe des Vol. 25 der Schriftenreihe „Advances in Biochemical Engineering". Springer Verlag, Berlin—Heidelberg, 1982.
Die Artikel von C.-M. Chang und G. T. Tsao (Affinity Chromatography) bilden eine Einheit mit gemeinsamem Literaturverzeichnis. Im ersten legen die Autoren die Prinzipien einiger Theorie dar, welche zur Behandlung affinitätschromatographischer Säulenprozesse geeignet sind. Mit den hier gegebenen Hinweisen auf die Originalliteratur ist dem Interessenten ein Einstieg möglich gemacht. Allerdings betonen die Autoren zu Recht, daß die Adsorptionsisothermen der sogenannten biospezifischen Adsorption, welche die Grundlage dieser Technik ist, überwiegend nichtlinear sind. Damit entfällt eine der wichtigsten Vereinfachungen, welche die allgemeinen theoretischen Ansätze mathematisch praktikabel machen, was u. a. dazu geführt hat, daß die Theorie solcher nichtlinearer chromatographischer Systeme noch weitgehend offen ist.
Im zweiten Artikel gehen die Verfasser mehr auf die praktische Seite der Affinitätschromatographie ein. In kurzen Abschnitten werden jeweils die Trägermaterialien und die Chemie ihrer Funktionalisierung zu bioaffinen Adsorbentien mit Spacern behandelt. Es folgt die Betrachtung der störenden bzw. sich überlagernden Faktoren (hydrophobe und ionische Effekte) und Betrachtungen zur Bedeutung und Bestimmung der Adsorbenskapazität sowie der Gleichgewichts- und Geschwindigkeitskonstanten des Vorgangs mit einem Ausblick auf Anwendungen.
Ein weiterer Artikel übermittelt neben einer nochmaligen gedrängten Erörterung der Grundprinzipien der Affinitätschromatographie eine sehr instruktive Übersicht über alle Probleme, die bei der Maßstabsvergrößerung auftreten. Dies erfolgt im besonderen Hinblick auf Proteine, doch gilt vieles hierzu gesagte allgemein. Wenn dieser Artikel auch niemals eigene experimentelle Erfahrungen ersetzen kann, so kann er doch sehr helfen, Vorabeinschätzungen bei praktischen Versuchen verschiedensten Maßstabs zu treffen, Irrwege zu vermeiden und Entwicklungsmöglichkeiten zu erkennen. Zugleich gibt er eine informative Übersicht über bereits getätigte quasitechnische Anwendungen der Proteinaffinitätschromatographie in Dimensionen, welche über die des Laboratoriums hinausgehen. Zugleich werden aber auch einige Ursachen deutlich, warum dies bisher noch Probleme bietet.
Im letzten Artikel geben die Verfasser, die selbst zu den Pionieren auf diesem Gebiet gehören, eine ausgezeichnete Übersicht über die Anwendung fixierter Triazinfarbstoffe, die sich zu einer sehr leistungsfähigen und breit anwendbaren Sondermethode der sog. Affinitätschromatographie herausgebildet hat. Nach einer kurzen historischen und chemischen Übersicht wird der Charakter der Farbstoff-Protein-Wechselwirkungen erörtert. Die Praxis der Farbstoffkupplung an den Träger wird detailliert beschrieben und es werden allgemeine methodische Aspekte sowie die Reinigung einiger ausgewählter Enzymgruppen besprochen. Am Schluß legen die Autoren die Grundgedanken und darauf fußende Weiterentwicklungsmöglichkeiten nochmals kurz dar.
Insgesamt liefert der sehr empfehlenswerte und instruktive Band einen vorzüglichen Überblick über die enorme Potenzen bietende Methode der biospezifischen Adsorption. Daß sich den genannten Grundkörpern in zunehmendem Maße synthetische Grundmatrizes an die Seite stellen werden, hätte vielleicht schon jetzt stärker betont werden können.

G. Langhammer

Acta Biotechnol. 5 (1985) 3, 245–250

Biodegradability of Swine Effluents

I. Oxygen Consumption*

OLIVEIRA, J. S., COSTA, M. L., RODRIGUES, A., MENEZES, J. G. A., PEREIRA, A. M. L.

Sanitary Engineering Section, Faculty of Sciences and Technology, New University of Lisbon, Portugal

Summary

Positive and negative effects of dilution of raw effluents from swine breeding are discussed, as to the treatability of those waste waters. Its effects on oxygen consumption and on the variation of inhibitory actions possibly observed, are studied and quantified. Using BRADENDER and VANDEPUTTE techniques based on peptone degradation, and determining PHELPS's equation parameters for different consumption curves observed, the advantages of dilution versus treatment of raw effluents is studied.

1. Introduction

The impact of swine breeding effluents on environmental quality attains very great importance in selected regions. In previous communications some of those problems have been discussed, including the treatment processes recommended and/or used in Portugal as well as the methods of upgrading effluents through biomass production.
The treatability of swine effluents by biological methods is a consequence of the biodegradability of the materials present in the wastewaters. But it is not the same, from an engineering point of view, to treat the effluents as they are produced or to dilute them, lowering the organic load and increasing the volume of wastewaters to be treated. The positive and negative effects of that option, its consequences on oxygen consumption and on the concentration of eventual inhibitors are analysed. The kinetics of bioremoval of pollutants will be discussed in a second paper.

2. Material and Methods

Swine effluents used in this study were collected from two different installations:

a) A small private lot located in Monte da Caparica, near the campus, 10 km from Lisbon, whose characteristics were presented elsewhere (OLIVEIRA et al., 1983)
b) The installations of the National Institute for Veterinary Research in Lisbon.
Analyses were made according to the methods indicated in [4] and to the following:

- Dissolved oxygen determined by the electrochemical probe method.
- Biochemical oxygen consumption determined by using an HACH system with 5 positions.
- ATP by using a Lumac Biocounter M 2010.

* This work was partially supported by Volkswagen Foundation under research contract.

3. Results and Discussion

Oxygen consumption during biodegradation of the organic matter present in the effluents is usually determined by using global analytical processes like biochemical oxygen demand (BOD).

This determination is rather controversial from a methodological and theoretical standpoint. The precision and reliability of BOD determination are quite reduced, and the relative importance of non-oxydative biodegradation reactions, the input of molecular oxygen released during the process and its consumption due to nitrification are variables whose control is unfeasible at the moment. Nevertheless, it is an index of microbial oxygen consumption under more or less standard conditions and represents the satisfaction of energy and carbon needs of heterotrophics during a closed biodegradation process. The limitations mentioned above do not allow the establishment of a quantitative and absolute relationship between oxygen consumption and biodegrated organic matter present.

Taking into account all these limitations, BOD is nevertheless an index of global organic pollution and can be used in that sense and also for the evaluation of the eventual presence of hazardous substances in the sample.

For our treatment scheme it is obviously important to study what happens to the oxygen consumption levels during biodegradation of swine effluents. The influence of biostatic or biocidic actions of substances present in the swine effluents is important for the project of treatment installations, determining different kinetic parameters and reactor dimensions.

In a first series of assays we used a respirometer according to HACH, to determine the oxygen consumption during several days (conditioned by a maximum level of O_2 of 300 mg/l) with the effluents indicated as in 2.a). The alternatives studied were:

a) Raw effluents
b) Dilution of 1 : 15 of raw effluents
c) Dilution of 1 : 20 of raw effluents
d) Dilution of 1 : 25 of raw effluents
e) Dilution of 1 : 30 of raw effluents

The values obtained up to the stabilization of the O_2 readings were used to adjust the PELP's equation

$$L_t = L_0 \, e^{-k't}$$

being

L_t — oxygen consumption at time t
L_0 — oxygen consumption on the 20th day of the assay
K' — oxydation coefficient of the substrate (base e)
t — incubation period in days

Using Thomas'methodology we calculated the parameters K' and L, for the different alternatives, whose values are presented in Table 1.

In a second series of assays, using the effluents indicated in b), we obtained the values for the same parameters which are presented in Table 2.

The alternatives studied in this case were:

a) Raw effluents
b) Dilution of 1 : 10 of raw effluents
c) Dilution of 1 : 20 of raw effluents
d) Dilution of 1 : 30 of raw effluents

Table 1. Values of K and L_0 for different dilutions of the same sample (Monte da Caparica)

Parameters	Dilution of the effluent				
	1 : 1	1 : 15	1 : 20	1 : 25	1 : 30
K (days^{-1})	0.374	0.282	0.295	0.330	0.250
L_0 (mg O_2 l^{-1})	568	470	438	397	359

Table 2. Values of K and L_0 for different dilutions of homogenized samples from National Institute for Veterinary Research (Lisbon)

Parameters	Dilution of the effluent			
	1 : 1	1 : 10	1 : 20	1 : 30
K (days^{-1})	0.366	0.374	0.390	0.350
L_0 (mg O_2 l^{-1})	463	406	350	253

Table 3. Values of K and L_0 for different dilutions of the sample from M. Caparica, supplement with peptone.

Parameters	Dilution of the effluents				
	1 : 1 + Peptone	1 : 15 + Peptone	1 : 20 + Peptone	1 : 25 + Peptone	1 : 30 + Peptone
K (days^{-1})	0.370	0.304	0.300	0.362	0.282
L_0 (mg O_2 l^{-1})	580	481	465	423	396

Table 4. Percentage of inhibition for different dilutions from the samples from M. Caparica.

Dilution	Inhibition [%]
1 : 1	−98.33
1 : 15	−90.43
1 : 20	−87.36
1 : 25	−86.52
1 : 30	−83.65

It is important to indicate that the total oxygen demand TOD values for the two samples are different. The sample a) presents an average value of 19500 mg O_2/liter and in sample b) the value was 7267 mg O_2/liter, reflecting different water managements in the exploitation and the type of breeding scheme.

The samples had previously been homogenized with ultra-sonic treatment to minimize the influence of heterogeneity in the results.

In Figs. 1 and 2 the consumption curves are presented. The influence of dilution is obvious.

Using the techniques of BRABENDER and VANDEPUTTE [1] applied to a BOD determination according to HACH apparatus, we studied the possible presence of inhibiting

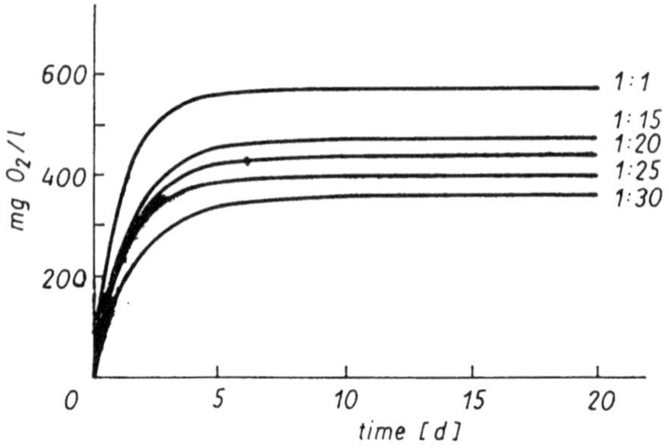

Fig. 1. Oxygen consumption in swine effluents with different dilution rates (from M. Caparica private lot).

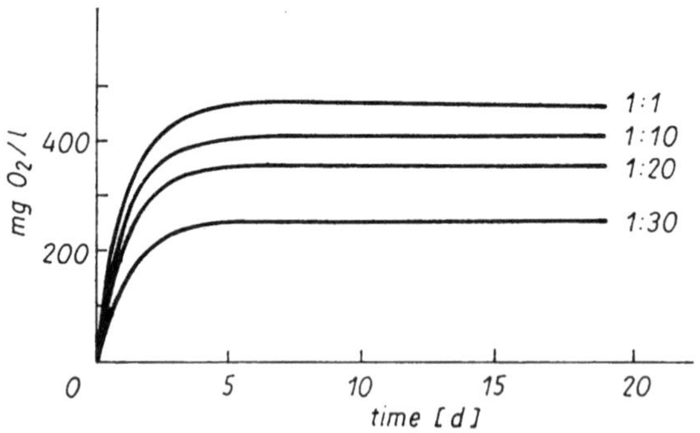

Fig. 2. Oxygen consumption in swine effluents with different dilution rates (from the National Institute for Veterinary Research)

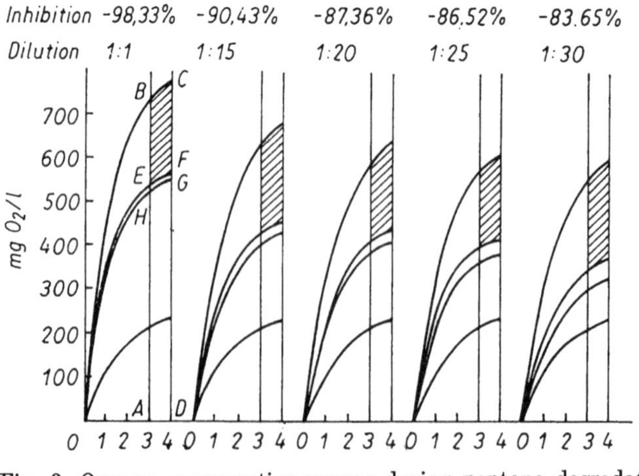

Fig. 3. Oxygen consumption curves during peptone degradation for the determination of inhibition in percentage.

substances in the effluents and the influence of dilution on the magnitude of the interference observed.
One of the explanations for the non-proportionality existing between dilution and O_2 consumption is obviously the concentration of the effluents and its effect upon microorganism growth. But the methodology referred to above allowed us to prove the existence of inhibition problems. In Fig. 3 the curves corresponding to the O_2 consumption are shown for the following situations:

1 — Raw effluents
2 — Peptone (28 mg N/l)*
3 — Raw effluents + Peptone
4 — Sum of curves 1) and 2)

The shaded zone in the stabilized phase of O_2 consumption is a measure of inhibition verified with the effluents studied. The percentage of inhibition is determined by adopting the expression:

$$\text{Inhibition} = \frac{\text{HEFG} \times 100}{\text{HBCG}} - 100 \qquad (1)$$

Using the effluents from the Monte da Caparica private lot the influence of peptone added to the effluents is obvious, increasing the values of K and L_0.
In Fig. 3 the curves corresponding to the different experimental alternatives are presented, and the inhibition is expressed according to equation (1) by the percentage of the area of the shaded zone in relation to the area comprised between curves 1 and 4.
The increase of dilution allowed for a more effective biodegradation of the organic material present in the effluents, reducing the inhibition effects.
The dilution alternative that will be adopted does not only influence the final quality of the effluents but also the type of treatment system that must be used. The concentration of refractory substances due to the inhibition processes will make it necessary to adopt more sophisticated systems to obtain an adequate quality of the effluents if that is demanded.

4. Conclusions

Biological depuration of swine effluents is possible. In the biodegradability assays presented the authors tried to study some of the processes of the removal of substances from the original substrate. Valorization of that substrate through biomass production is a possibility already studied to complement and to make feasible from an economic standpoint of this treatment.
The major problem that we have considered in the context of this research is to balance the positive and negative consequences of dilution of the effluents during biodegradation. Concerning substances present in the effluents with biostatic and/or biocidic effects, dilution presents advantages, having a higher removal efficiency and reducing inhibitory effects. In a second part of this paper the biokinetics of pollutants removal and its consequences for treatment systems and their efficiency will be analyzed.

Received June 25, 1984

* The consumption of O_2 in the peptone solution (28 g N l^{-1}) is expressed by the equation:

$$L_t = 259.4 \times 10^{-0.233t}$$

References

[1] BRABANDER, K., VANDEPUTTE, H.: Cebedeau **329** (1971), 188.
[2] GASPAR, N., GASPAR, A., OLIVEIRA, J. S.: Characterisation and treatment of swine effluents. Symp. Environ. Manag. Devel. Countries, Istanbul. 1984.
[3] OLIVEIRA, J. S.: Operacões e Processos Fundamentais em Engenharia Sanitária. Fundamentos de depuracão biologica. UNL, Lisboa, 1982. P. 256.
[4] OLIVEIRA, J. S. et al.: Wissenschaft und Umwelt **4** (1983), 252.
[5] REMICHE, C., WIELEN, V. D.: Cebedeau **381/382** (1975), 314.
[6] ROQUES, H.: Fondaments théoriques du traitement biologique des eaux. vol. 2. Technique et Documentation, Paris, 1979. P. 1813.

Book Review

EBERT/EDERER

Computeranwendungen in der Chemie

Eine Einführung in das Arbeiten mit Kleinrechnern

Weinheim: Verlag Chemie, 1983. 359 S., 63 DM.

Dem Buch liegt die Programmsprache BASIC zugrunde — eine Sprache, die leicht zu erlernen ist und die sich bei der Nutzung von Kleincomputern durchgesetzt hat. Anhand von Aufgaben, wie sie sich dem Chemiker, aber nicht nur ihm, stellen, werden Anweisungen der Programmsprache und die benötigte Mathematik erläutert. Die Anwendungsbeispiele bauen aufeinander auf, so daß die behandelten Programme zunehmend komplexer werden — sie sind aber in jedem Falle vollständig abgedruckt, wodurch auch derjenige das Buch nutzen kann, der sich nur über die Bewältigung einzelner Aufgaben und das zugehörige Programm unterrichten will. Darüberhinaus gibt es zu jeder Aufgabe, die erklärt und mit Hilfe des abgedruckten Programmes gelöst wird, ergänzende Aufgaben, die der Leser selbst lösen soll.
Die Autoren beginnen anhand des Programmierens von einfachen Formeln mit einer verständlichen Einführung in BASIC. Im weiteren werden Beispiele behandelt zu: Berechnung von Reihen durch Iteration, Numerische Integration, lineare und nichtlineare Gleichungssysteme, Differentialgleichungen und Interpolationsverfahren. Ein alphabetisches Programmverzeichnis sowie ein alphabetisches Verzeichnis der BASIC-Sprachelemente runden das Buch ab.
Es vermittelt in methodisch guter Weise Kenntnisse, eigene Programme zu schreiben für die Lösung von Problemen wie die Auswertung von Meßergebnissen, die Berechnung von Parametern oder die Konstruktion von Diagrammen und Ausgleichskurven. Damit ist es ein wertvoller Helfer sowohl für Studierende als auch für Praktiker in Laboratorien und Technik, deren Tätigkeitsgebiet in der Chemie aber auch angrenzenden Fachgebieten, einschließlich der Biotechnologie, liegt.

Chr. Albrecht

Acta Biotechnol. 5 (1985) 3, 251–261

Biodegradability of Swine Effluents

II. Kinetics of Pollutants Removal*

OLIVEIRA, J. S., COSTA, M. L., RODRIGUES, A., MENEZES, J. G. A., BORIÉ, M. C.

Sanitary Engineering Section
Faculty of Sciences and Technology
New University of Lisbon, Portugal

Summary

Some aspects of the kinetics of bioremoval of pollutants during biodegradation of swine-breeding waste waters are studied and a biphasic model is proposed. The concentration of nitrogen and phosphorus compounds during the process is followed. Following a previous contribution on the same subject, the aspects of dilution of those effluents are discussed, and it is proposed to treat raw effluents, in spite of the reduction of possible biostatic actions. Dilution will be acceptable only if strong inhibitory effects are observed.

1. Introduction

The biodegradability and biodegradation kinetics of swine breeding effluents are important aspects of the treatability of these hazardous waste waters. In a previous paper we have already discussed the oxygen consumption during the biological depuration and the interference of some inhibitory substances in the process [14].

Continuing the analysis of positive and negative effects of dilution, in this context, we shall discuss some aspects of the kinetics of the bioremoval of pollutants during biodegradation in this paper. The removal of organic substances present in the effluents will be considered separately from biomass evolution, due to the different nature of these two aspects of the process: substrate consumed and living agents of consumption. The evolution of nitrogen and phosphorus compounds in the analytical fractions considered will be followed and the biochemical meaning of those transformations will be discussed.

The consequences of dilution for treatment systems as well as their design and operation are the main purposes of this study.

2. Material and Methods

Using the effluents whose origin and characterization had been presented elsewhere [14], analyses were made following procedures indicated in OLIVEIRA et al. [13] and also of:

— Total oxygen demand (TOD), using an apparatus from Philips, model PW 9625

* This work was partially supported by Volkswagen Foundation under research contract

- Total carbon and total organic carbon (TOC) determined by rapid combustion, using an apparatus from the Oceanography International Corporation, model 525, and measuring the CO_2 released by infrared analysis with an Horibe spectrophotometer, model PIR-2000
- ATP, using a LUMAC Biocounter M 2010

3. Biodegradation of the effluents

3.1. Removal of organic substrate

To rationalize the design criteria of reactors, a study of the biodegradation of these effluents in batch reactors was made, and models of the results were established.
One of the subjects studied was the removal of organic carbonaceous pollutants present in the medium.
In Tables 1 and 2 experimental results obtained in two typical assays are presented, using raw swine effluents as substrate and the dilution of 1 : 10 of the effluents.
The heterogeneity of the product made it necessary to repeat the biodegradation assays several times.
Six assays were made with the raw effluents and eight experiments were performed with the 1 : 10 dilution.
Samples were collected at selected aeration periods and several parameters were determined, to control the system evolution.

Table 1. Biodegradability of raw effluents from swine breeding

Parameters	Aeration time (days)								
	0	0.2	1.0	3.0	6.0	9.0	13.0	15.0	20.0
Total oxygen demand ($mg\ O_2\ l^{-1}$)	4320	4187	3920	3167	2700	2233	2140	n.d.	2113
Total carbon ($mg\ C\ l^{-1}$)	752	784	710	524	420	384	242	n.d.	236
Conductivity (μ Si)	5500	5750	6025	5750	5500	5575	5250	5250	4500
pH	8.56	8.65	8.62	8.62	8.73	8.65	9.05	9.00	8.87
rH (mV)	+54	n.d.	−34	−62	−115	−252	+58	+15	+23
Total suspended solids ($mg\ l^{-1}$)	3270	3990	4470	4600	3530	1490	810	n.d.	114
Volatile suspended solids ($mg\ l^{-1}$)	2210	3250	3710	3730	3210	1100	690	n.d.	82
CO_2 ($mg\ l^{-1}$)	0	0	0	0	0	0	0	n.d.	0
Alkalinity (Methylo red)	190	170	200	n.d.	n.d.	n.d.	n.d.	n.d.	n.d.
($mg\ CaCO_3\ l^{-1}$) (Phenolphthaleine)	1680	1920	2030	n.d.	n.d.	n.d.	n.d.	n.d.	n.d.
Total − N ($mg\ N\ l^{-1}$)	840	785	770	653	653	558	530	n.d.	357
NH_4^+ − N ($mg\ N\ l^{-1}$)	487.2	554.4	548.4	420.0	459.2	476.0	437.0	n.d.	297.0
N − NO_2^- ($mg\ N\ l^{-1}$)	2.27	3.44	3.30	0.42	0.31	0.08	0.03	n.d.	0.03
N − NO_3^- ($mg\ N\ l^{-1}$)	68.0	79.3	40.3	12.2	10.0	7.0	7.5	n.d.	5.9
Orthophosphate $mg\ P\ l^{-1}$)	94.5	n.d.	98.4	114.0	137.7	45.7	26.3	n.d.	12.6

Table 2. Biodegradation of a diluted effluent (1 : 10) from swine breeding

Parameters	Aeration time (days)								
	0	0.2	1.0	3.0	6.0	9.0	13.0	15.0	20.0
Total oxygen demand (mg O_2 l^{-1})	833	737	600	350	273	223	212	n.d.	200
Total carbon (mg C l^{-1})	94	64	104	120	114	94	86	n.d.	89
Conductivity (μ Si)	950	975	1025	1000	1025	1050	1000	1080	1040
pH	8.26	8.34	8.33	8.17	8.27	8.40	8.50	8.41	8.46
rH (mV)	+84	+82	+41	+74	+104	+80	+84	+31	+68
Total suspended solids (mg l^{-1})	345	460	790	550	n.d.	80	38	n.d.	48
Volatile suspended solids (mg l^{-1})	250	242	410	460	n.d.	70	34	n.d.	16
CO_2 (mg l^{-1})	0	0	0	0	0	0	0	0	0
Alkalinity (Methylo red)	10	30	30	30	30	35	40	n.d.	40
(mg $CaCO_3$ l^{-1}) (Phenolphthaleine)	90	270	280	310	350	340	340	n.d.	260
Total — N (mg N l^{-1})	123	106	101	85	70	67	70	n.d.	57
NH_4^+ (mg N l^{-1})	59.4	72.8	78.4	72.8	67.2	50.4	67.0	n.d.	540
NO_2^- — N (mg N l^{-1})	0.15	0.16	0.19	0.12	0.11	0.01	0.01	n.d.	0.02
NO_3^- — N (mg N l^{-1})	5.32	4.58	3.58	2.56	4.22	2.91	2.18	n.d.	1.45
Orthophosphate (mg P l^{-1})	11.3	18:2	10.8	13.7	12.4	5.2	n.d	n.d.	5.1

Pollutant load control was made by TOD determinations. The individual variability of the samples made it necessary to transform them, using the percentage value, in relation to the initial one (time 0, days = 100). With these indices the evolution along the aeration period (20 days) was calculated and numerical values computed; using as initial reading the average of the values obtained with raw and diluted effluents. Table 3 shows the results obtained.

Table 3. Removal of pollutant load from raw and diluted effluents, expressed in TOD (mg O_2 l^{-1})

Aeration time (days)	Raw effluent		Diluted effluent (1 : 10)	
	Index	Average calculated value	Index	Average calculated value
0.0	1.00	3502	1.00	883
0.2	0.97	3390	1.03	913
1.0	0.96	3362	0.58	514
3.0	0.72	2532	0.40	355
6.0	0.70	2451	0.33	295
9.0	0.58	2035	0.28	244
11.0	0.53	1863	—	—
13.0	—	—	0.26	228
18.0	—	—	0.22	195
20.0	0.51	1800	0.21	187

The model proposed by ECKENFELDER and FORD [4] and by BENEDEK et al. [2] was then applied to these values (Fig. 1A and B). It corresponds to removal kinetics of biodegradable volatile suspended solids of the first order according to the expression

$$\frac{(X_d)_e}{(X_d)_0} = e^{-K_b t}$$

where

$(X_d)_e$ — degradable volatile suspended solids (VSS) remaining after batch aeration time t, expressed in mg l^{-1}
$(X_d)_0$ — initial degradable VSS at time 0, expressed in mg l^{-1}
K_b — batch reaction rate of the destruction of degradable VSS, expressed in days^{-1}
t — time of aeration, expressed in days.

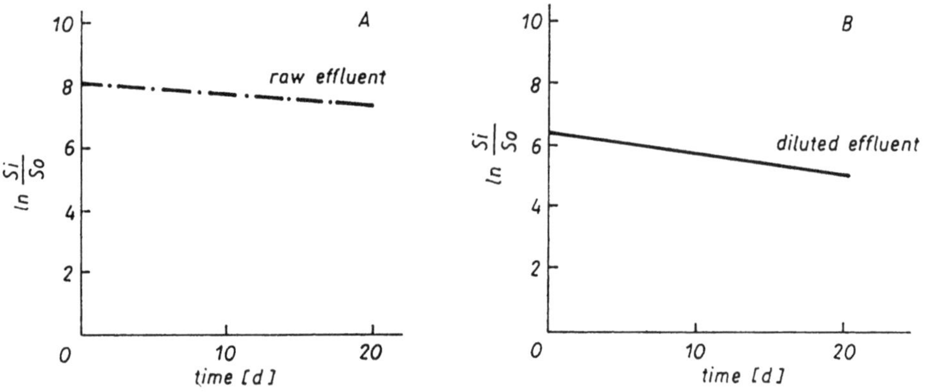

Fig. 1. Removal of biodegradable solids in batch reactor according to the model proposed by ECKENFELDER and FORD (1970).
A — Raw effluents, B — Diluted effluents (1 : 10)

Using raw effluents, the removal can be expressed by

$$\frac{(X_d)_e}{(X_d)_0} = 614.5\, e^{-0.0702t}$$

with $r^2 = 0.7753$, $F = 9$ and a standard deviation equal to 174.1. For the diluted effluents the expression obtained was

$$\frac{(X_d)_e}{(X_d)_0} = 3176.0\, e^{-0.0361t}$$

with $r^2 = 0.8187$, F = 24 and a standard deviation equal to 308.7.

The general applicability is acceptable, but, as predicted, the model of ADAMS et al. [1] gave much better results. In this model we take into account the degradable solids (X_d) expressed as a function of initial VSS (X_0), correcting the values with non-degradable residue (X_n) admitted constantly during the whole period of aeration

$$\frac{X_e - X_n}{X_0 - X_n} = e^{-K_b t} \qquad \text{(Fig.2)}$$

For the raw-effluents case the expression obtained is

$$\frac{X_e - 1800}{X_0 - 1800} = 1.173\, e^{-0.271t}$$

being $r^2 = 0.9174$, $F = 58$ and $s = 0.127$.

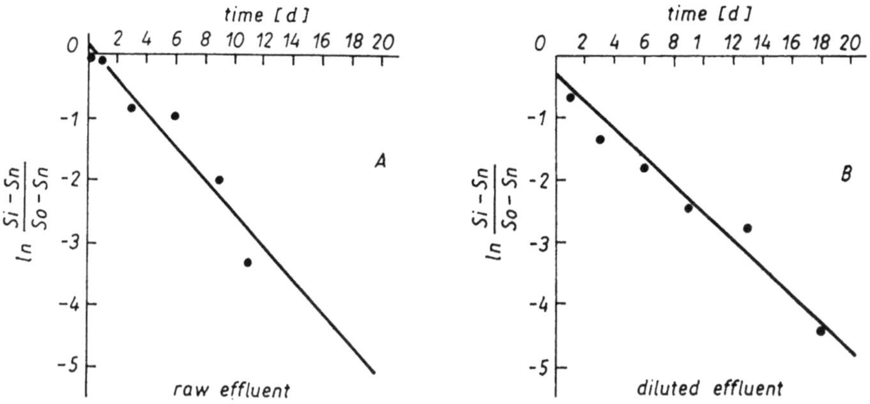

Fig. 2. Removal of biodegradable solids in batch reactor according to the model proposed by ADAMS et al. (1974).
A — Raw effluents, B — Diluted effluents (1 : 10)

For the diluted effluents we have

$$\frac{X_e - 187}{X_e - 187} = 0.725\ \mathrm{e}^{-0.221t}$$

with $r^2 = 0.9534$, $F = 10$ and $s = 0.202$.
Even with this good adjustment we have tried a better fit with the model proposed by TUCEK, CHÚDOBA and MADERA [18], using the value of K_b obtained with the exponential model initially proposed.
Utilising the same indices proposed by the authors

$$\Pi 1 = \frac{\Delta S}{S_0}$$

$$\Pi 2 = K_b Z$$

$$Z = \text{retention time (days)}$$

we obtain the following equations to model the removal of organic carbonaceous pollutants (Fig. 3 A and B).

a) Raw effluents ($K_b = 0.0362$)

$$\Pi 2/\Pi 1 = 1.685\ 2\Pi + 0.848$$

being $r^2 = 0.9702$, $s^2 = 0.079$ and $F = 97$

b) Diluted effluents ($K_b = 0.0702$)

$$\Pi 2/\Pi 1 = 1.201\ \Pi 2 + 0.110$$

being $r^2 = 0.9991$, $s^2 = 0.019$ and $F = 5841$.

The results obtained have a high level of significance. The biological basis of removal processes observed indicated that eventually the MICHAELIS-MENTEN model according

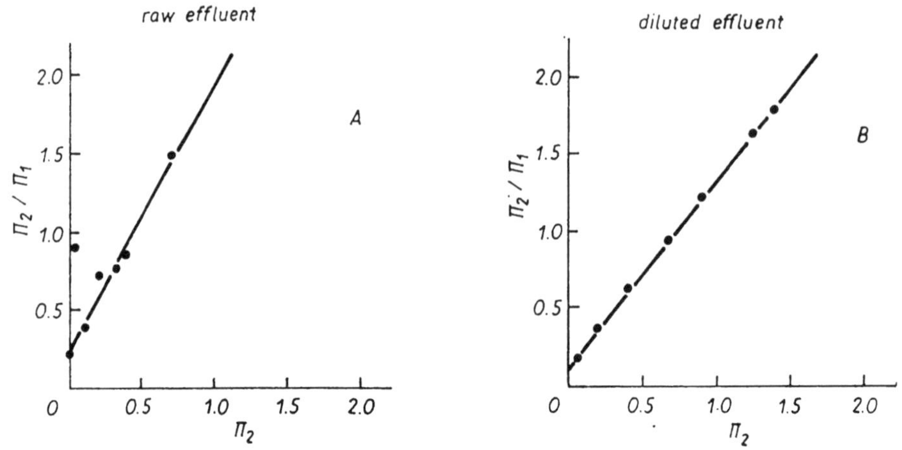

Fig. 3. Removal of organic carbonaceous pollutants in swine effluents according to the model proposed by TUCEK, CHUDOBA and MADERA (1971).
A — Raw effluents, B — Diluted effluents (1 : 10)

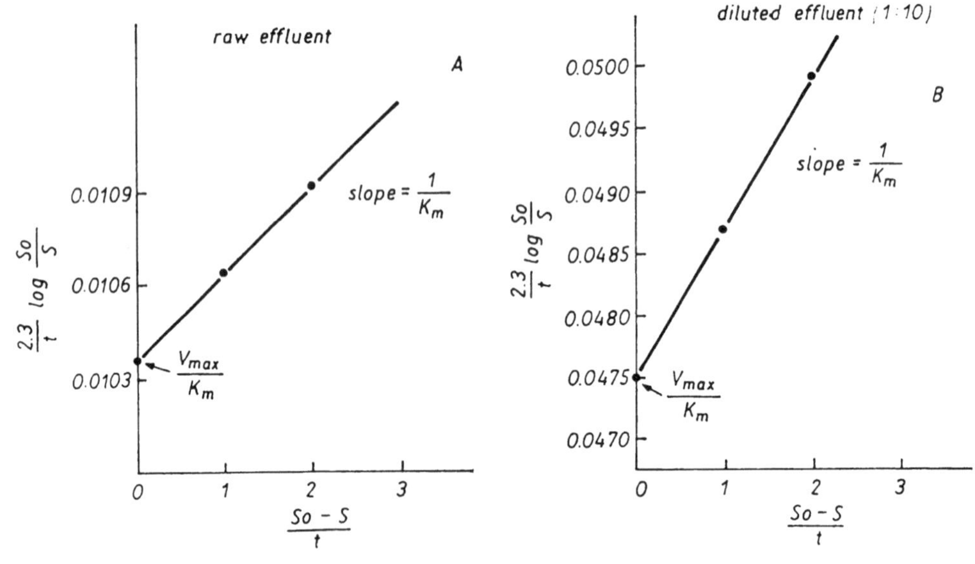

Fig. 4. Integrated form of MICHAELIS-MENTEN model according to MONOD's formulation to swine effluents.

A Raw effluents

$r^2 = 0.9748$	$K_m = 3571.4$ mg
$S = 0.006$	$V_{max} = 40.0\ d^{-1}$
$F = 970$	

B — Diluted effluents (1 : 10)

$r^2 = 0.8943$	$K_m = 833.3$ mg
$S = 0.047$	$V_{max} = 39.6\ mg\ l^{-1}\ d^{-1}$
$F = 296$	

to MONOD's formulation could have been used. Using the integrated equation, the following results were obtained (Fig. 4 A and B)

a) Raw effluents (25 degrees of freedom)

$K_m = 3571$ mg $\quad V_{\max} = 40.0$ mg l^{-1} days^{-1}

with $r^2 = 0.9748$ and $F = 970$ (highly significant)

b) Diluted effluents (37 degrees of freedom)

$K_m = 833$ mg $\quad V_{\max} = 39.6$ mg l^{-1} days^{-1}

being $r^2 = 0.8943$ and $F = 296$ (highly significant).

As predicted, the enzymatic system shows a reduced affinity to the substrate with evident inhibitory processes, as we had already detected in the oxygen consumption kinetics.

3.2. Biomass evolution

The quantity of biomass present in the reactor was controlled during the aeration period by several methods

- Total organic carbon
- Volatile suspended solids
- ATP

For the raw-effluents case the biomass evolution is presented in Table 4, utilizing percentage indices of actual values in relation to the original reading to express average biomass in the different assays.
The initial increase of biomass corresponds to the consumption of released nutrients as a consequence of the oxydative processes occurring in the reactor (phase of dominant synthesis). When nutrients are consumed the system enters into a phase of dominant endogenous respiration, with lysis of cell material. Using VSS as a biomass control parameter, the contribution of organic colloids, without biological activity but associated with living matter, is obvious. The use of the TOC test is obviously a better index of biological biomass but the presence of non-living carbonaceous organics is also evident, mainly in the residual substances at the end of the aeration period.

Table 4. Indices of biomass evolution in the reactor

Aeration time (days)	Raw effluent		Diluted effluent (1 : 10)		
	TC	VSS	TOC	VSS	ATP
	$\bar{x} = 714$ mg	$\bar{x} = 2425$ mg	$\bar{x} = 306$ mg	$\bar{x} = 187$ mg	$\bar{x} = 1111$ pg/ml
0.0	1.00	1.00	1.00	1.00	1.00
0.2	1.03	1.27	0.97	1.35	1.37
1.0	0.98	1.38	0.84	1.50	9.69
3.0	0.82	1.51	0.57	1.71	—
6.0	0.67	1.16	0.59	—	0.31
9.0	0.46	0.54	0.54	0.30	0.16
11.0	—	—	—	0.16	0.30
13.0	0.32	0.19	0.53	—	0.12
18.0	—	—	0.36	—	—

The evolution of biomass in diluted effluents is similar in its general trends to the kinetics verified with raw effluents. After a dominant synthesis phase, in the beginning, the dominant endogenous respiration period after about 6 days of aeration is evident. This point was followed in greater detail is this case using measures of ATP as an index of living matter. The increase in cell numbers during the first day of aeration is extremely important. The loss of the sample on the third day made it impossible to fix the inversion of the evolution more precisely in this case. Nevertheless, the alteration of observed values obeys the classical standards of population evolution in a closed system.
The modeling of biomass evolution made it necessary to make a previous choice of the analytical method to be applied in its control, as different tests have different biochemical meanings. Using VSS, the traditional parameter utilized with this type of reactors, we adjusted the global evolution in two steps:

a) In the phase where $VSS_i/VSS_0 \geqq 1$ (dominant synthesis phase) we adjusted a second degree polynomial curve.
b) When $VSS_i/VSS_0 \leqq 1$ (dominant endogenous respiration phase) the evolution follows an exponential model

For the first phase the results obtained were the following:

1) Raw effluents (17 experimental values)

$$X = 1 + 0.1923t - 0.0293t^2$$

being X biomass expressed in VSS present at the moment t

($r^2 = 0.961$, $F = 36.4$, significant)

2) Diluted effluents (13 experimental values)

$$X = 1 + 0.3449t - 0.0489t^2$$

Being $r^2 = 0.961$, $F = 53.0$ (highly significant)

For the second phase, being X the concentration of biomass present (mg l^{-1}) and t the aeration time, the degradation follows an exponential model

1) Raw effluents

$X = 3.129\ e^{-0,201t}$ ($r^2 = 0.9854$, $F = 56$)

with $P \geqq 99.9\%$

2) Diluted effluents

$X = 4.163\ e^{-0.295t}$ ($r^2 = 0.9997$, $F = 10413$)

with $P \geqq 99.9\%$

The first phase dominates up to the 6th day and the second phase after the 6th day of aeration.
The transition between these phases depends obviously on the characteristic of the effluents, on the population and generally on the conditions prevailing during biodegradation in the system (Fig. 5A and B).

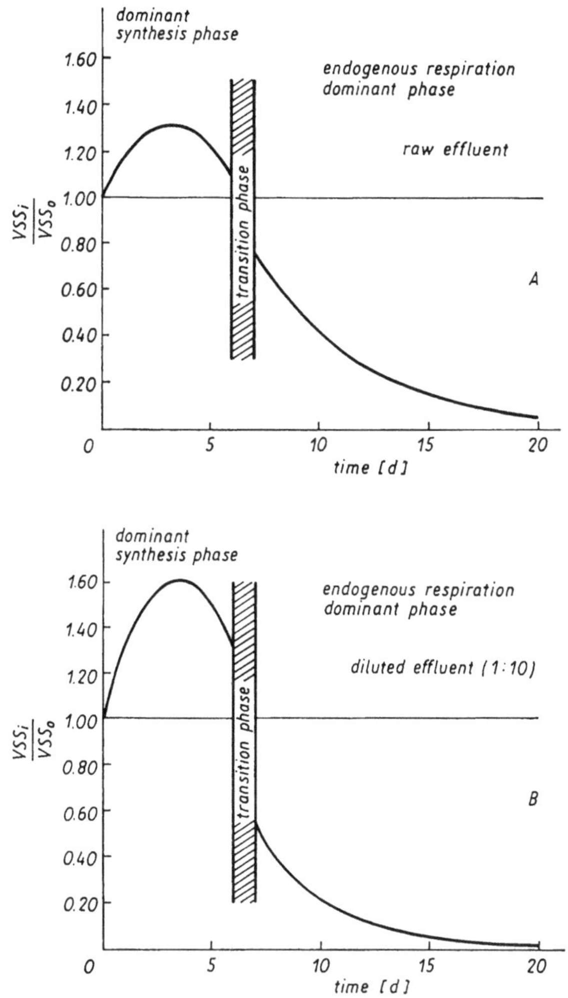

Fig. 5. Model proposed for biomass evolution during biological treatment of swine effluents in a batch reactor.
A — Raw effluents (1 :10), B — Diluted effluents (1 :10)

3.3. Evolution of nitrogen and phosphorus compounds

The evolution of nitrogenous substances is presented in Table 5 and Fig. 6. It corresponds to the usual biodegradation cycle, with an initial hydrolysis of proteins following oxidative reactions in the system, increasing the concentration of $N-NH_4^+$. The transition from synthesis to hydrolysis in the reactor, induces some apparent irregularities in the general trend of the processes, finishing at a value of about 60% of initial $N-NH_4^+$ content due to fixation in the biomass and loss to the atmosphere as NH_3, by aeration and mixing the liquid. The evolution is similar in raw and in diluted effluents.
The oxidized nitrogenous forms show a general tendency to decrease which is a reflection of the non-development of nitrifying bacteria, probably due to the inhibitory effect of ammonia.

Table 5. Evolution of nitrogenous and phosphorous compounds during biodegradation period

Aeration time (days)	Raw effluent					Dilluted effluent (1 : 10)				
	N-total	N—NH_4	N—NO_3	N—NO_2	P—H_2PO_4	N-total	N—NH_4	N—NO_3	N—NO_2	P—H_2PO_4
0	818	503	68.0	2.18	86.1	131.5	60.0	6.70	0.22	14.0
0.2	813	529	75.6	2.26	n.d.	106.0	72.8	6.10	0.22	18.1
1	792	560	41.2	3.60	108.4	102.5	75.6	3.58	0.28	14.8
3	670	412	15.3	0.56	121.0	85.9	70.0	2.93	0.15	16.8
6	663	468	9.8	0.39	125.7	72.0	64.4	4.22	0.26	15.6
9	538	479	6.8	0.06	53.2	72.5	49.0	3.49	0.01	5.5
13	516	454	7.8	0.04	31.2	73.0	57.3	3.20	0.02	3.6
20	323	294	5.8	0.04	12.8	48.6	43.4	1.82	0.02	5.6

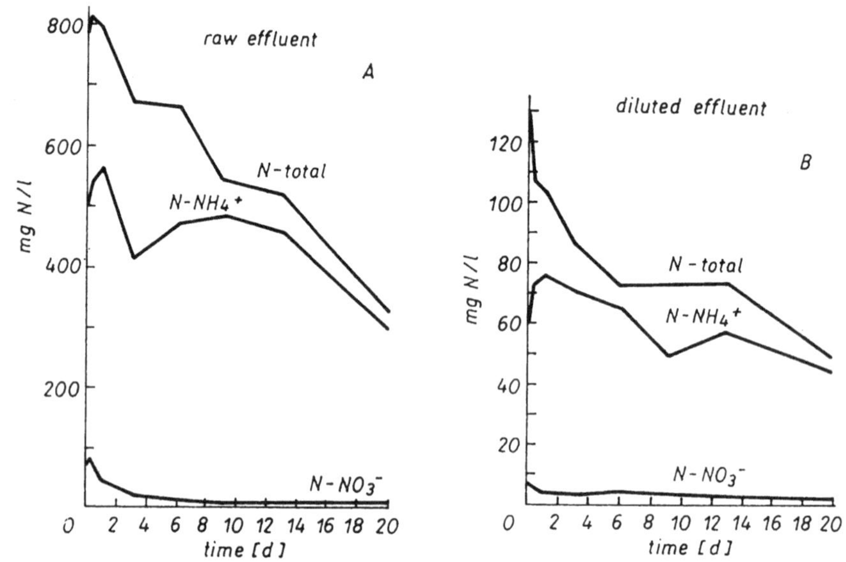

Fig. 6. Evolution of nitrogenous fractions during biological biodegradation of swine effluents in a batch reactor.
A — Raw effluents, B — Diluted effluents (1 : 10)

The evolution of orthophosphates also presented in Table 5 corresponds to the dominance of oxidative mechanisms and also of chemical precipitation. The mineralization of organic matter under aerobic conditions releases cations immobilized in those substances in the original situation. When solubility products of the different inorganic phosphates are attained, there is a removal of orthophosphates present in the soluble phase.

On the other hand, there is a bioaccumulation of phosphorus in living organisms and in the biological flocs with of reduction of the P-content of the liquid phase.

4. Conclusions

Swine breeding effluents make an important impact upon the environmental quality of many rural areas, as do other animal production waste waters. The biodegradability of those substances makes the biological depuration of the effluents possible.

The dilution of waste waters makes the treatment easier, reducing the concentration and the activity of the inhibitory or at least biostatic pollutants present.
But the increase in volume to be treated drives up the costs in the operation and maintenance of those facilities without significant savings. So, the effluents produced during swine breeding in conditions similar to those where our samples were obtained can be treated without dilution. If exogenous substances are added during the process, and those substances have a significant biocidic and/or biostatic effect during biological depuration, it could be necessary to dilute the effluents in order to solve treatment problems. In other cases, the economic benefits obtained by raw effluents treatment are superior to the technological advantages gained by treating diluted effluents in the depuration system.

Received June 25, 1984

References

[1] ADAMS, C., ECKENFELDER Jr., W. W., STEIN, R.: Water Res. **8** (1974), 213.
[2] BENEDEK, P., FARKAS, P., LITERATHY, P.: Water Res. **6** (1972), 91.
[3] DAIGGER, G., GRADY Jr., C. L.: Water Res. **16** (1982), 365.
[4] ECKENFELDER Jr., W. W., FORD, D. L.: Water Pollution Control. Austin: The Pemberton Press, Texas. 1970.
[5] ESENER, A. A. et al.: Biotechnol. Bioeng. **23** (1981), 2391.
[6] GATES, W., MARLAR, J.: J. Water Pollut. Control Fed. **40** (11, Part 2) (1968), R. 469.
[7] GOSH, S., POHLAND, F.: Water Res. **6** (1972), 99.
[8] HERBERT, D., ELSWORTH, R., TELLING, R. C.: J. Gen. Microbiol. **14** (1956), 619.
[9] KIM, J. W., HUMENICK, M., ARMSTRONG, N.: Water Res. **15** (1981), 1221.
[10] KIM, J. W., ARMSTRONG, N.: Water Res. **15** (1981), 1233.
[11] KIM, J. W., ARMSTRONG, N.: Water Res. **15** (1981), 1249.
[12] LAWRANCE, A., MCCARTY, P.: J. Sanit. Eng. Div.
[13] OLIVEIRA, J. S. et al.: Wissenschaft und Umwelt **4** (1983), 252.
[14] OLIVEIRA, J. S. et al.: Acta Biotechnol. **5** (1985) 3, 233.
[15] ROQUES, H. et al.: Water Res. **16** (1982), 839.
[16] SEGEL, I. H.: Bioquimia. Rio Janeiro, Livros Tecn. e. Cient. Ed. SA. 1979. 572.
[17] TAIGANIDES, E. P.: Bioengineering properties feed lot wastes. — In: Animal Wastes. Ed.: TAIGANIDES, E. P. London: Applied Science Publ., London, 1977. 131—151.
[18] TUCEK, F., CHUDOBA, J., MADERA, V.: Water Res. **5** (1971), 647.
[19] VAVILIN, F. A.: Biotechnol. Bioeng. **24** (1982), 1921.

Acta Biotechnol. 5 (1985) 3, 262

Book Reviews

U. Jensen; D. E. Fairbrothers

Proteins and Nucleic Acids in Plant Systematics

Berlin, Heidelberg, New York, Tokyo: Springer-Verlag, 1983. 408 pp., 148 Fig., DM 114

In the last decade research on proteins and nucleic acids in plants and their role and function has led to new insights and informations. The use of this knowledge in plant physiological, systematical and phytogenetical research is opening new fields of view. In this volume these fields are summarized. It contains papers presented at an international symposium, entitled „Proteins and nucleic acids in plant systematics" held at the university of Bayreuth/FRG in July 1982.
In five sections about — Nucleic acids; Proteins: structural properties; Particular proteins contributing to phytogeny and taxonomy; Serological protein properties contributing to phytogeny and taxonomy; Serological data and current plant classification — some well-known experts summarize the knowledge in the special fields. The papers are written in a clear and informative manner. The numerous figures and tables give a concise illustration of the topics of this symposium.
The bibliographic data are listed up to 1983. In a conclusional paper the editors and D. Boulter give a critical review of the knowledge and its possible using in plant systematics. This volume is an important part in the series "Proceedings in Life Sciences" and it is excellent in equipment, print and clearness. This volume is obliged to have a fixed place in all relevant libraries and laboraties.

H.-P. Schmauder

Trevor A. Kletz

Cheaper, safer plants or wealth and safety at work — Notes on inherently safer and simpler plants

Loos Prevention
Rugby/England: The Institution of Chemical Engineers, 1984.

118 S., £ 8

Moderne Chemieanlagen erfordern auf Grund ihrer Größe und der komplizierten Technologien aufwendige Sicherheitssysteme. Trevor A. Kletz versteht sein Buch als Beitrag zur Senkung des Sicherheitsrisikos in Anlagen vor allem der chemischen Industrie. Sein Konzept der inhärenten Sicherheit besteht darin, zum frühest möglichen Zeitpunkt bei der Entwicklung des Verfahrens die sicherheitstechnischen Erfordernisse zu berücksichtigen und durch eine entsprechend sichere Gestaltung der Technologie Gefährdungen zu vermeiden oder zu beschränken. Er gibt Anregungen, Hinweise und diskutiert Beispiele, wie sich die Sicherheit von Anlagen ohne zusätzlichen Aufwand an Sicherheitstechnik verbessern läßt.
Das Buch ist an den Bedürfnissen der Praxis orientiert und ist für Ingenieure der Verfahrensentwicklung und Projektierung von besonderem Interesse. Es ist allgemeinverständlich geschrieben und enthält eine Vielzahl aussagekräftiger Abbildungen. Es empfiehlt sich als Lehrmaterial zur Einbeziehung in die Vorlesungen und Seminare der Fächer Projektierung und Sicherheitstechnik an Hochschulen und Universitäten, besonders für die Studienrichtung Verfahrenstechnik.

H. Wolke

Acta Biotechnol. 5 (1985) 3, 263–269

Stereoselective Hydrolysis of Optically Active Isomeric Isopulegyl Acetates by Rhodotorula mucilaginosa and Bacillus sp.

MONKIEWICZ, A. and GÓRA, J.

Institute of General Food Chemistry
Technical University of Łódź
Stefanowskiego 4/10, 90–924 Łódź

Summary

A method for isolation of optically pure *l*-isopulegol from a mixture of its optically active isomers using the microorganisms *Rhodotorula mucilaginosa* and *Bacillus* sp. is described.
Microorganisms hydrolyzed *l*-isopulegyl acetate (26–40%) and in a small degree d-isopulegol acetate whereas the *d*-neoisopulegol acetate remained non-hydrolyzed.
The optical purity of the chromatographic pure *l*-isopulegol was 97.6%.

Biochemical methods suitable for resolution of racemic compounds are actively investigated and several terpenoids of industrial interest have been resolved recently by the use of microorganisms [1–4].
One of the technical methods of synthesis of *l*-menthol is based on the acid catalyzed cyclization of *d*-citronellal which provides a mixture of isomeric isopulegols consisting of *l*-isopulegol (72%), *d*-neoisopulegol (23%), *d*-isoisopulegol (5%) and traces of *d*-neoisoisopulegol [5]. *l*-Isopulegol is then usually isolated from this mixture by freezing

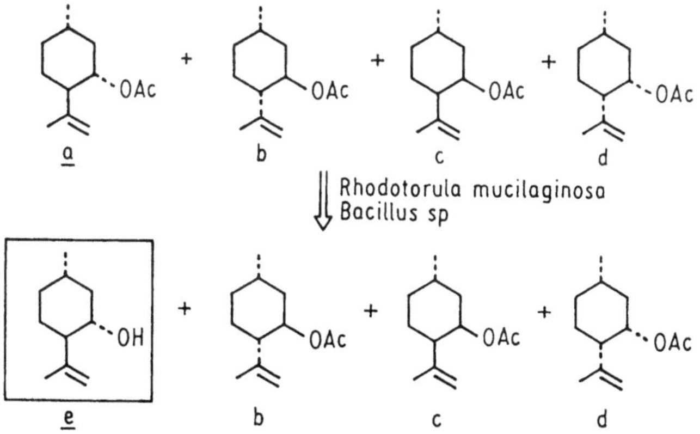

Fig. 1. Hydrolysis of isomeric isopulegyl acetates by microorganisms
a — *l*-isopulegyl acetate
b — *d*-isoisopulegyl acetate
c — *d*-neoisopulegyl acetate
d — neoisoisopulegyl acetate
e — *l*-isopulegol

out at temperature — 60 °C and its subsequent hydrogenation over Raney Ni-catalyst gives the desired *l*-menthol [5]. The isolation of *l*-isopulegol is crucial to this sequence. Currently, yields of about 50% can only be achieved in this step.
Our recent success [6] in the separation of *l*-menthol from the racemic mixture of menthyl acetates via enantioselective microbial hydrolysis utilizing *Rhodotorula mucilaginosa* and *Bacillus* sp. prompted us to use these microorganisms for the analogous isolation of *l*-isopulegol from the mixture of its isomers mentioned above. In this paper we wish to present our results which demonstrate that both *Rhodotorula mucilaginosa* and *Bacillis* sp. can be utilized in the selective hydrolysis of isomeric isopulegyl acetates leading to the separation of about 97% optically pure *l*-isopulegol in a satisfactory yield (Fig. 1).

Materials and Methods

Substrates

A mixture of optically active isopulegols ("Herbapol") containing 74% of *l*-isopulegol, 24% of *d*-neoisopulegol and 2% of *d*-isoisopulegol was transformed into a mixture of the corresponding acetates by treatment with acetic anhydrideanhydrous sodium acetate. The obtained mixture of isopulegyl acetates was purified by distillation under reduced pressure.
B.p. 72–74 °C/2 Torr; n_D^{20} 1.4576; $[\alpha]_D^{20}$ — 0.565° (neat);
IR: 1735, 1235 and 1025 cm^{-1}.

Biological Material

Rhodotorula mucilaginosa R_{10} and *Bacillus* sp. S-2 used for stereoselective hydrolysis of isopulegy acetates were obtained from the Pure Culture Collection of the Fermentation Technology and Microbiology Institute of the Technical University of Łódź.

Culture Media

Media of the following compositions were used for cultivation

— for yeast

1. medium A (yeast extract 0.5; glucose 2.0; peptone 0.7; and K_2HPO_4 0.5%).
2. medium B (yeast extract 0.5; glucose 0.5; peptone 0.5; K_2HPO_4 0.5; and agar-agar 0.2%).

— for bacteria

1. medium C (meat extract).
2. medium D [2] (peptone 1.0; meat extract 0.4; glucose 0.1; and agar-agar 0.2%).
3. medium E [7] (beef extract 0.33; peptone 1.11; and NaCl 0.56%).

Cultivation Conditions

Microorganisms were cultivated periodically with shaking at 28 °C in 30 ml Erlenmeyer flasks containing 30 or 40 ml of the appropriate medium or, in 500 ml flasks containing 30 ml of medium. Shaking frequencies of 170 and 250 strokes/min, respectively, were applied.
Inoculum of *Rhodotorula mucilaginosa* was cultivated for 24 h and then it was added in 1 ml portions to a number of flasks containing medium of same composition. The cultivation was then carried out for 17 to 48 h. The resting cells of *Rhodotorula mucilaginosa* were harvested after 48 h of cultivation, washed twice with 0.1 M potassium phosphate buffer (pH 7.0) and then suspended in 40 ml of this buffer. Cultivation of *Bacillus* sp. was carried out for 12 to 48 h. Growth parameters of the microorganisms were estimated in the usual way [8, 9].

Hydrolysis Conditions

The mixture of isopulegyl acetates was added to the culture solutions in 0.5—3.0% portions (per volume of the medium or of the phosphate buffer). Shaking was then applied for 24—72 h; reaction temperature was kept at 28 °C during the whole process.

Isolation and Analysis of Hydrolysis Products

Reaction products consisting of *l*-isopulegol and unreacted isomeric isopulegyl acetates were isolated from the culture broth by steam distillation followed by extraction of the distillate with n-hexane. The hexane extracts were dried over anhydrous $MgSO_4$ and evaporated. The oily residue was analyzed by GC using a 3 m steel column packed with OV-17.3% on Gas — Chrom Q at temperature 100—220 °C with gradient 12 deg./min and flow rate of nitrogen 40 ml/min or, on a 2 m steel column packed with Apiezon L on 5% Chromosorb W (AW-DMCS) at temperature 110—200 °C with 8 deg./min and flow rate of nitrogen 40 ml/min. The content of *l*-isopulegol in this residue is called "the hydrolysis ratio" further in the text.
l-Isopulegol was separated from the unreacted isopulegyl acetates by its quantitative esterification with boric acid in toluene, with azeotropic removal of water [10]. Toluene and isopulegyl acetates were then distilled off and the remaining *l*-isopulegyl orthoborate was hydrolyzed during the steam distillation. *l*-Isopulegol was extracted from the distillate with n-hexane and the hexane extracts were then combined and dried over anhydrous $MgSO_4$. Evaporation of the solvent afforded crude *l*-isopulegol which was purified by Kugelrohr distillation (b.p. 110 °C/12 Torr), chromatographed on Kieselgel 60 (230—400 mesh, chloroform-acetone) and finally distillated again (Kugelrohr). *l*-Isopulegol of the same optical purity was separated from the non-hydrolyzed isopulegyl acetates in more useful way involving column chromatography on Kieselgel 60 (230—400 mesh, chloroform-acetone) and Kugelrohr distillation. Optical rotations were measured on a Perkin-Elmer spectrophotometer at 20 °C and the specific rotation, $[\alpha]^{20}_{546.1}$ — 25.1° (c = 5.1, EtOH) was obtained for *l*-isopulegol which corresponds to a 97.6% optical purity of the sample [11].

Results and Discussion

Growth Parameters of Rhodotorula mucilaginosa and Bacillus sp.

Growth parameters of *Rhodotorula mucilaginosa* obtained during its cultivation on medium A were compared with those obtained on medium B. Growth rate and biomass level of the periodical yeast cultures on the standard medium A usually exceeded those observed on the medium B (Table 1). Growth curves determined on both culture media are typical for a two-stage process (in case of medium B the second stage was more distinctive) (Fig. 2). Such a situation might have originated from the different rates of consumption of medium components.
At the early stage of the cultivation of yeast, intensive uptake of a readily accessible energy source, e.g. glucose, would lead to the intensive growth of *Rhodotorula mucilaginosa* on the media rich in glucose, and consumption of the other non-sugar components of the medium would then occur in the second stage of its growth.

Table 1. Growth Parameters of *Rhodotorula mucilaginosa* and *Bacillus* sp.

Strain	Culture medium	Specific growth rate μ [h^{-1}]	Generation time T [h]	Maximum biomass production [mg/ml]
Rhodotorula	A	0.216	3.2	4.80
mucilaginosa	B	0.111	6.2	0.65
Bacillus sp.	C	0.358	1.9	0.50

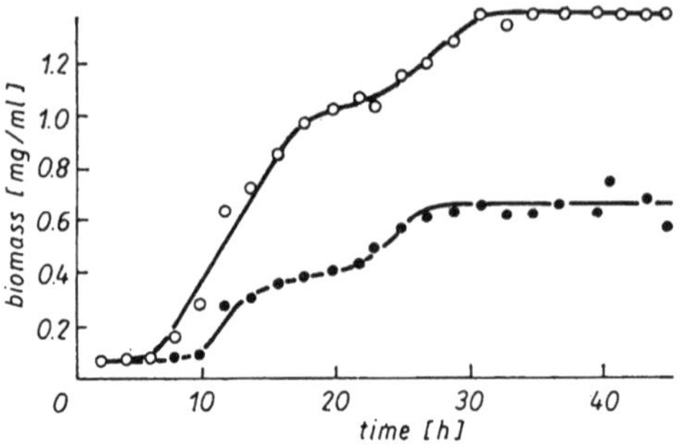

Fig. 2. Growth curves of *Rhodotorula mucilaginosa*
○ on standard medium A
● on medium B

On the other hand, such a two-stage appearance of the growth curves might be associated with a reversible formation of cell aggregates at different stages of the growth of yeast. On the both culture media *Rhodotorula mucilaginosa* attained the maximum of the biomass level after 30 h of cultivation.
The growth of *Bacillus* sp. on meat extract medium occured typically and the bacteria reached its stationary growth phase after 15 h of cultivation (Fig. 3).

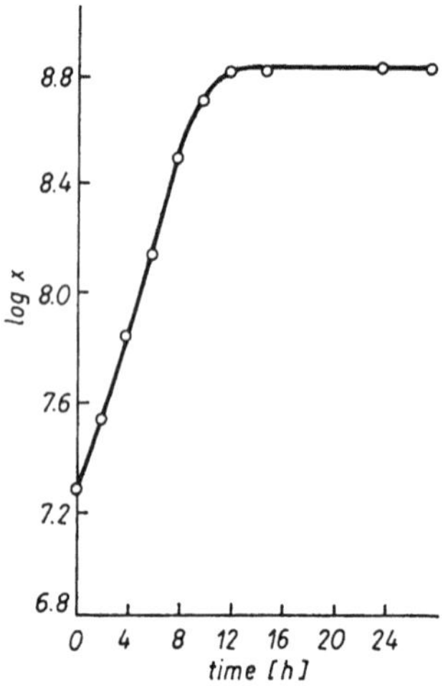

Fig. 3. Growth curve of *Bacillus* sp.

Hydrolysis of Isopulegyl Acetates by Rhodotorula mucilaginosa

Results of the hydrolysis of isopulegyl acetates by the growing and resting cells of *Rhodotorula mucilaginosa* are presented in Table 2. The data indicate clearly that the growing cells of *Rhodotorula mucilaginosa* were more active against *l*-isopulegyl acetate on the medium B rather then on the medium A, despite the fact, that this latter medium was found to be optimal for its growth.
It can therefore be suggested that hydrolytic activity of *Rhodotorula mucilaginosa* should not be connected only with the biomass level and the rate of the growth (presented in Table 1). The other factors like substrate concentration, hydrolysis time or metabolic state of the cells should be also taken into consideration. Thus, populations of growing cells of *Rhodotorula mucilaginosa* were found to be most active during all their stationary growth phase, while the cells from the logarithmic growth phase showed markedly lowered activity.

Table 2. Results of Stereospecific Hydrolysis of Isopulegyl Acetates by *Rhodotorula mucilaginosa*

No	Growth phase[a]	Isopulegyl acetates concentration % [v/v]	Culture medium[b]	Product ratio[c] [%] I	II
1.	advanced stationary	0.2	A[d]	9.3	90.7
2.	advanced stationary	1	A[d]	2.0	98.0
3.	advanced stationary	2	A	8.5	91.5
4.	early stationary	2	A	3.5	96.5
5.	advanced stationary	0.5	B	40.0	60.0
6.	advanced stationary	1	B	26.3	73.7
7.	early stationary	0.5	B[e]	40.1	59.9
8.	early stationary	1.5	B[e]	16.1	83.9
9.	logarithmic	1	B[e,f]	5.3	94.7
10.	advanced stationary	0.5	B[d]	31.0	69.0
11.	advanced stationary	3	B[d]	6.2	93.8
12.	advanced stationary	1	B[h]	30.9	69.1
13.	advanced stationary	0.7	B[h]	32.3	67.7
14.	advanced stationary	0.7[g]	A[h]	35.1	64.9

[a] — time of cultivation — growth phase
17 h — logarithmic
29 h — early stationary
48 h — advanced stationary

[b] — 30 ml of culture medium; shaking frequency 180 strokes/min; time of hydrolysis 72 h

[c] — *l*-isopulegol I: unreacted acetates II (e.g. *l*-isopulegyl acetate + *d*-neoisopulegyl acetate + *d*-isoisopulegyl acetate) as determined by GC analysis; averaged values calculated from 10 runs

[d] — shaking frequency 250 strokes/min

[e] — 40 ml of culture medium

[f] — isopulegyl acetates added gradually in small portions

[g] — time of hydrolysis 140 h

[h] — resting cells suspended in phosphate buffer solution

The yield of hydrolysis was found to be strongly concentration dependent. At the concentration of the mixture of isomeric isopulegyl acetates below 1% (v/v) *l*-isopulegyl acetate was hydrolyzed by the growing yeast cells with the hydrolysis ratio of 40%. However, an increase of the concentration of isopulegyl acetates to 1% caused the hydrolysis ratio to drop to 26,3%.

Interestingly, at higher concentrations of isopulegyl acetates the resting cells of *Rhodotorula mucilaginosa* harvested on medium A and on medium B were found to be more

effective in hydrolysis of *l*-isopulegyl acetate than the corresponding growing cells of the same biomass content. However, in this case, lowering of the concentration below 1% was not accompanied by any meaningful improvement of the yield.
After isolation of *l*-isopulegol the unreacted isopulegyl acetates were usually recycled. In no case the presence of *d*-neoisopulegol, the possible byproduct resulting from the hydrolysis of *d*-neoisopulegyl acetate was detected.

Hydrolysis of Isopulegyl Acetates by Bacillus sp.

Our results concerning hydrolysis of isopulegyl acetates by the use of *Bacillus* sp. are collected in Table 3. As can be seen from the data in Table 3, *Bacillus* sp. exhibited pronounced activity against *l*-isopulegyl acetate under variety of reaction conditions. It was found to be most effective on meat extract medium C. Use of isopulegyl acetates in concentration of 0.5% (v/v) gave the best results and lead to the formation of *l*-isopulegol with the hydrolysis ratio of 40.2% which corresponded to 60.3 μl of this substrate in a 30 ml batch. It seemed more practical to us, however, to work at the concentration level of 3% because as much as 203 μl of *l*-isopulegol could be produced in one

Table 3. Stereospecific Hydrolysis of Isopulegyl Acetates by *Bacillus* sp.

No	Growth[a] phase	Isopulegyl acetates concentration % [v/v]	Culture medium[b]	Product ratio[c] [%]	
				I	II
1.	late stationary	0.5	C	40.2	59.8
2.	late stationary	1	C	35.7	64.3
3.	late stationary	2	C[d]	23.0	77.0
4.	late stationary	1	E	29.7	70.3
5.	late stationary	1	D	14.4	85.6
6.	advanced stationary	1	C[e]	12.9	87.1
7.	advanced stationary	1	C	17.4	82.6
8.	advanced stationary	1	E[f]	22.9	77.1
9.	advanced stationary	1	D[f]	23.6	72.4
10.	advanced stationary	1	C[f]	30.5	69.5
11.	early stationary	0.5	C	27.9	72.1
12.	early stationary	2	C	9.7	90.3
13.	early stationary	3	C	7.8	92.2
14.	late stationary	3	C	22.6	77.4

[a] — cultivation time — growth phase
12 h — early stationary
19 h — stationary
24 h — advanced stationary
48 h — late stationary
[b] — 30 ml of culture medium; time of hydrolysis 48 h
[c] — C.f. footnote c, Table 2
[d] — time of hydrolysis 72 h
[e] — time of hydrolysis 24 h
[f] — bacteria stored on medium D — agar slopes

fermentation run (in 30 ml batch), although the hydrolysis ratio decreased to 22.6% in such cases.

Bacillus sp. harvested at the late stage of its stationary growth phase turned out to be more active than that collected at the early stage of this phase. Additionally, considerable decrease in the hydrolytic activity of this latter biological material upon increase of concentration of isopulegyl acetates was also noted.

In our hydrolysis experiments utilizing *Bacillus* sp. the isolated *l*-isopulegol was of the same 97% optical purity as the one obtained by the use of *Rhodotorula mucilaginosa*. Again, the unreacted isopulegyl acetates were recycled. Under the conditions applied *d*-neoisopulegyl acetate remained always non-hydrolyzed.

In summary, it should be mentioned that the presented results compare well with the observations by ORITANI and YAMASHITA [2] who have similarly employed several moulds and bacteria for the separation of *l*-isopulegol from the mixture of its isomers obtained by cyclization of *dl*-citronellal. Reported hydrolysis ratios for *l*-isopulegyl acetate were 45.2; 38.6; and 35.2% (by GC) for *Trichoderma S*, *Bacillus subtilis var Niger*, and *Absidia hyalospora* growing cells, respectively. *Rhodotorula mucilaginosa* was not studied by these authors.

It should be pointed out that the presented investigation of microbial hydrolysis of isomeric isopulegyl acetates is meant preliminary to the elaboration of this process for the large scale preparation of *l*-isopulegol in a fermenter. A two-step procedure involving freezing out of *l*-isopulegol from a mixture of isomeric isopulegols followed by acylation of the residue and subsequent microbial hydrolysis of isomeric isopulegyl acetates seems to be especially attractive in this respect.

The authors acknowledge financial support for this work by a grant MR II-17 from University of Łódź.

Received November 4, 1983

References

[1] MONKIEWICZ, A., GÓRA, J.: Pollena TSPK **26** (1982), 40.
[2] ORITANI, T., YAMASHITA, K.: Agric. Biol. Chem. **37** (1973), 1691.
[3] ORITANI, T., YAMASHITA, K.: Agric. Biol. Chem. **37** (1973), 1687.
[4] ORITANI, T., YAMASHITA, K.: Agric. Biol. Chem. **38** (1974), 1961.
[5] KULESZA, J., GÓRA, J., KOWALSKA, K., DOGIELSKA, Z., DRURI, M.: Przemysł Chemiczny **50** (1971), 571.
[6] OBERMAN, H., GÓRA, J., MONKIEWICZ, A., FLORIANOWCZ, T.: Adv. Biotechnol. **3** (1981), 513.
[7] MORI, K., AKAO, H.: Tetrahedron **36** (1980), 91.
[8] PIRT, J. S.: Principles of Microbe and Cell Cultivation. Ch. 3. Oxford: Blackwell Scientific Publications, 1975. 15.
[9] KOTEŁKO, K., SEDLACEK, L., LACHOWICZ, T. M.: Biologia bakterii, PWN, Warszawa 1977. 284.
[10] GÓRA, J., OBERMAN, H., MONKIEWICZ, W.: Pl-PS 221847 (1980).
[11] PICKARD, R. H., HUNTER, H., LEWCOCK, W., de PENNINGTON, H. S.: J. Chem. Soc. (London) **117** (1920), 1248.

Acta Biotechnol. 5 (1985) 3, 270

Book Review

Paul Präve, Uwe Faust, Wolfgang Sittig, Dieter A. Sukatsch

Handbuch der Biotechnologie

2. Auflage/Studienausgabe

München, Wien: R. Oldenbourg Verlag 1984. 616 S., 197 Abb., 84 DM

In der zweiten Auflage erscheint das Handbuch der Biotechnologie hier als preisgünstige Studienausgabe. In diesem Werk geben Wissenschaftler aus Forschung und Praxis einen Überblick über den Wissensstand aller an der Biotechnologie beteiligten Fachgebiete. Die verfahrenstechnischen Vorgänge werden ebenso beschrieben, wie die dafür nötigen Grundlagen.
Das Handbuch richtet sich nicht nur an den erfahrenen Praktiker und Spezialisten, sondern besonders auch an den Anfänger, der, aus dem Studiengang einer Fachrichtung kommend, die anderen Disziplinen kennenlernen muß. Biotechnologie ist eine im hohen Maße interdisziplinäre Wissenschaft.

Aus dem Inhalt —
Einleitung:

Geschichte, Verfahren und Produkte.

Biotechnologische Grundlagen:

Mikroorganismen; Stoffwechsel der Mikroorganismen, Mikrobieller Sekundärstoffwechsel; Pflanzliche und tierische Zellkulturen; Bioreaktoren; Biologische Regulation und Prozeßführung; Aufarbeitung in der Biotechnologie.

Verfahren und Produkte:

Gärungsverfahren; Primärmetabolite; Enzyme; Antibiotika, Biotransformatoren; Mikrobielle Biomassegewinnung; Umweltbiotechnologie; Mikrobielle Biomassegewinnung; Laugung (Leaching); Wirtschaftliche Aspekte.

Anhang:

Zeitschriften, Periodika, Reviews; Abkürzungen, Formeln und Symbole; Gesetze, Richtlinien und Vorschriften; Arbeitsgrundlagen; Herstellende Firmen; Produktliste.
Sachwortverzeichnis.

Acta Biotechnol. 5 (1985) 3, 271–278

Modellierung der enzymatischen Cellobiosehydrolyse — Vergleich linearer und nichtlinearer Parameterbestimmung

BAUER, F.[1] und KUDE, J.[2]

[1] Akademie der Wissenschaften der DDR
Zentralinstitut für Isotopen- und Strahlenforschung
DDR-7050 Leipzig, Permoserstraße 15
[2] Akademie der Wissenschaften der DDR
Institut für Biotechnologie
DDR-7050 Leipzig, Permoserstraße 15

Summary

Experimental kinetic data (initial rate and high conversion) on the hydrolysis of cellobiose by 1,4-β-glucosidase (*Gliocladium* sp.) have been analysed and a competitive inhibition by glucose has been proposed. The determination of kinetic parameters from integral data is based upon algorithms for non-linear optimization and numerical integration. The values of kinetic constants ($v_{max} = 1.02 \frac{\mu M_{glucose}}{mg_{protein} \cdot min}$, $K_M = 2.6$ mM/l, and $K_P = 1.2$ mM/l) agree well with the initial-rate results. An important distinction is the confidence limit of parameters. Linear regression analysis shows a virtual accuracy and can lead to wrong conclusions.

Einleitung

Die Rechtfertigung und die Notwendigkeit der Einführung der mathematischen Modellierung in die Biotechnologie ist ausführlich von PÖHLAND [1] diskutiert worden. Bei der Erstellung eines mathematischen Modells für einen Prozeß werden neben Wärme- und Stofftransportübergangsgrößen auch Kenntnisse über den Reaktionsmechanismus und die entsprechenden kinetischen Parameter benötigt. Die Bestimmung dieser unbekannten Parameterwerte aus experimentellen Daten und die Selektion zwischen mehreren Mechanismusvorstellungen ist ein komplexes mathematisches Problem. Historisch bedingt finden zur Ermittlung der kinetischen Konstanten unterschiedliche Linearisierungen der Geschwindigkeitsgleichungen Anwendung, wobei die dabei möglichen Fehler oft ignoriert werden. Trotz steigenden Interesses nimmt die Verwendung von Konzentration-Zeit-Kurven zur Parameterbestimmung von enzymkatalysierten Reaktionen nur einen geringen Umfang ein. Die Vorteile dieser als „nichtlinear" zu bezeichnenden Vorgehensweise bestehen in einer besseren Ausnutzung der in dem Konzentrationsverlauf enthaltenen Information (Reduzierung von Versuchsanzahl und Material) und in dem Umgehen der Ermittlung von Reaktionsgeschwindigkeiten (z. B. schwierig bei lag-Phasen in der Produktbildung). Weiterhin bringt die Behandlung von Mehrschrittreaktionen keine prinzipiellen Probleme mit sich, was für die linearisierten Geschwindigkeitsgleichungen nur in wenigen Fällen möglich ist. Als Einschränkung der Auswertung von Umsatz-Zeit-Kurven wird der Fall der Enzyminaktivierung angeführt. Das hierfür erforderliche Differentialgleichungssystem kann nicht mehr analytisch integriert werden, iterative Integrationsverfahren mit entsprechendem rechentechnischen Aufwand sind erforderlich. Hat man sich jedoch für die Versuchsauswertung mittels EDV-Anlage entschieden, so eröffnet sich auch eine weitere mathematische Durchdringung der Versuchsauswertung. Neben den Verfahren der nichtlinearen Regression zur Parameterermittlung stehen in den meisten Rechenzentren Methoden der Versuchsplanung zur Verfügung, die genauere Konstantenwerte und/oder eindeutigere Aussagen hinsichtlich des Reaktionsmechanismus als Ergebnis haben und bei heterogen-katalysierten Reaktionen Anwendung gefunden haben [2].

Theorie

In der Enzymkinetik hat sich die lineare Regression in Form der Lineweaver-Burk-Auftragung ($1/v_0$ vs. $1/S_0$) zur Gewinnung der kinetischen Konstanten bewährt. Weit weniger Anwendung finden die Verfahren von Hofstee (v_0 vs. v_0/S_0) und Hanes (S_0/v_0 vs. S_0). Eisenthal und Cornish-Bowden [3] bringen statistische und praktische Einwände gegen die obigen Auswerteverfahren vor. Die von ihnen vorgeschlagene Methode liefert insofern genauere Werte für v_{max} und K_M, da Ausreißer mit weniger Gewicht in die Parameterbestimmung eingehen. Günstig wirkt sich auch die Wichtung der in die lineare Ausgleichsrechnung eingehenden Werte aus. So zeigte Wilkinson [4], daß bei Annahme eines konstanten (bzw. relativen) experimentellen Fehlers die Abweichungen in der Lineweaver-Burk-Auftragung mit dem Faktor v^4 (bzw. v^4/S^2) zu multiplizieren sind.
Mit der Verwendung der integrierten Michaelis-Menten-Gleichung vermeidet man die Unzulänglichkeiten aus mathematisch-statistischer Sicht. Zur Auswertung gelangen die Konzentration-Zeit-Kurven, wobei die Umsatzgrade normalerweise hoch sind. Reversibilität, Produktinhibierung und Enzyminaktivierung kennzeichnen den Kurvenverlauf entscheidend. Die erhaltenen Geschwindigkeitskonstanten beschreiben von vornherein den gesamten Umsatzbereich und müssen nicht wie die unter Anfangsbedingungen gewonnenen Konstanten zusätzlich überprüft werden. Die in dieser Arbeit angewandte Strategie beinhaltet

— die Erstellung unterschiedlicher Modellansätze (basierend auf verschiedenen Reaktionsmechanismen)
— die numerische Integration der Differentialgleichungssysteme innerhalb der Parameterbestimmung mittels nichtlinearer Regression, wobei die Anfangswerte der Integration mit als Variable aufgefaßt werden
— die Auswahl des besten Modells auf Grund der Anpassung der berechneten Konzentrationsverläufe an die experimentellen Daten.

Unter folgenden Mechanismusvorstellungen für den Cellobioseabbau (S = Cellobiose, P = Glucose, E = 1,4-β-Glucosidase) sollte der wahrscheinlichste ermittelt werden.

<u>Modell 1</u> $E + S \underset{}{\overset{K_M}{\rightleftharpoons}} ES \xrightarrow{k} E + 2P$

$$-\frac{d[S]}{dt} = \frac{k \cdot \frac{1}{K_m} \cdot [E_0] \cdot [S]}{1 + \frac{1}{K_m} \cdot [S]}$$

<u>Modell 2</u> $E + S \overset{K_M}{\rightleftharpoons} ES \xrightarrow{k} E + 2P$

$E + P \overset{K_P}{\rightleftharpoons} EP$

$$-\frac{d[S]}{dt} = \frac{k \cdot \frac{1}{K_m} [E_0] [S]}{1 + \frac{1}{K_m} [S] + \frac{1}{K_P} [P]}$$

Modell 3

$$E + S \underset{}{\overset{K_M}{\rightleftharpoons}} ES \overset{k}{\rightarrow} E + 2P$$
$$E + P \overset{K_P}{\rightleftharpoons} EP$$
$$ES + P \overset{K_P}{\rightleftharpoons} ESP \overset{K_M}{\rightleftharpoons} EP + S$$

$$-\frac{d[S]}{dt} = \frac{k \cdot \frac{1}{K_M} \cdot [E_0][S]}{1 + \frac{1}{K_M} \cdot [S] + \frac{1}{K_P} \cdot [P] + \frac{1}{K_M \cdot K_P} \cdot [S][P]}$$

Modell 4

$$E + S \overset{K_M}{\rightleftharpoons} ES \overset{k}{\rightarrow} E + 2P$$
$$E + P \overset{K_P}{\rightleftharpoons} EP$$
$$ES + P \overset{K_P}{\rightleftharpoons} ESP \overset{K_M}{\rightleftharpoons} EP + S$$
$$ESP \overset{k}{\searrow} EP + 2P$$

$$-\frac{d[S]}{dt} = \frac{k \cdot [E_0] \left(\frac{1}{K_M} \cdot [S] + \frac{1}{K_M \cdot K_P} \cdot [S][P]\right)}{1 + \frac{1}{K_M} [S] + \frac{1}{K_P} \cdot [P] + \frac{1}{K_M \cdot K_P} \cdot [S][P]}$$

Modell 5

$$E + S \overset{K_M}{\rightleftharpoons} ES \overset{k}{\rightarrow} E + 2P$$
$$E + P \overset{K_P}{\rightleftharpoons} EP$$
$$E \overset{k'}{\rightarrow} E'_{\text{inaktiv}}$$

$$-\frac{d[S]}{dt} = \frac{k \cdot \frac{1}{K_M} \cdot [E_0] \cdot e^{-k' \cdot t} \cdot [S]}{1 + \frac{1}{K_M} \cdot [S] + \frac{1}{K_P} \cdot [P]}$$

Das Differentialgleichungssystem wurde mit einem RUNGE-KUTTA-Verfahren (4. Ordnung) integriert und die Konstantenbestimmung erfolgte durch Parameterschätzverfahren nach MARQUARDT [5] bzw. FLETCHER/POWELL [6]. In diesen iterativen Suchverfahren werden von Schätzwerten ausgehend die Parameterwerte solange geändert, bis für die vorgegebene Zielfunktion $Z(k)$ ein Minimum bei der Anpassung der theoretischen ($f_i(t_j)$) an die gemessenen (c_{ij}) Konzentrationswerte erreicht wird. Neben dem bekannten Fehlerquadratminimum als Zielfunktion der Optimierung

$$Z(k) = \sum_i \sum_j w_{ij} \cdot (c_{ij} - f_i(t_j, k, K_M, K_P))^2 = \text{Min!}$$

gelangte auch das von BOX [7] vorgeschlagene Kriterium

$$Z'(k) = \det \left| \sum_l (c_{il} - f_i(t_l))\,(c_{jl} - f_j(t_l)) \right| = \text{Min!}$$

zur Anwendung, da die Wahl geeigneter Gewichtsfaktoren w_{ij} infolge mangelnder Kenntnis der Standardabweichungen der Konzentrationen in gewissem Maße willkürlich wird. Der Meßfehler in den einzelnen Punkten wurde entweder als absolut konstant ($w_{ij} = 1$) oder relativ konstant ($w_{ij} = c_{ij}^{-2}$) angesehen. Die Wahl der Startwerte wurde durch die Ergebnisse der linearen Regressionsanalyse wesentlich erleichtert. Die individuellen Konfidenzbereiche der optimierten Parameterwerte k_i können infolge der Nichtlinearität des Optimierungsproblems nur näherungsweise abgeschätzt werden.

$$k = \hat{k} \pm t_{1-\alpha} \cdot \sqrt{\left(\frac{\partial^2 Z(k)}{\partial k_i\, \partial k_j}\right)^{-1} \cdot \frac{Z(\hat{k})}{(n-p)}}$$

wobei $t_{1-\alpha}$ der entsprechende Wert des t-Testes mit der Irrtumswahrscheinlichkeit α und $(n - p)$ der entsprechende Freiheitsgrad sind. Die Genauigkeit der erhaltenen Parameterwerte steigt erwartungsgemäß mit der Zahl der experimentellen Daten, daher sollten möglichst alle Reaktionspartner in die Versuchsauswertung einfließen. Sind die

Konstanten aller Modelle bestimmt, hat man in der minimierten Fehlerquadratsumme ein Maß für die Güte des Modells. Mit dem F-Test wird das beste Modell erhalten.

$$\frac{Z(k)_{\text{Modell } i}}{Z(k)_{\text{Modell } j}} = F_{\alpha}(n_i - p_i, n_j - p_j)$$

Es kann unter Umständen zwischen mehreren Modellansätzen mit dem vorliegenden Datenmaterial nicht unterschieden werden und ein gezieltes Vorgehen bei weiteren Versuchen macht sich erforderlich. Neben diesen globalen, statistischen Auswahlkriterien müssen physikalisch-chemische Gesichtspunkte bei der Modellselektion beachtet werden (z. B. positive Konstanten, zu erwartende Temperaturabhängigkeit, Eingliederung in Struktur-Reaktivität-Beziehungen). Betreffs weiterer Details bei der Integration der Differentialgleichungen und der Vorgehensweise in der Parameterschätzung sei auf die Arbeit von VOTRUBA [8] verwiesen.

Materialien

Als Enzymquelle der 1,4-β-Glucosidase (EC.3.2.1.21) wurde ein Cellulasepräparat von *Gliocladium* sp. mit einem Proteingehalt von 28% verwendet. Die Proteinkonzentrationsmessung erfolgte nach LOWRY et al. [9], wobei Pepsin vom Schwein als Standard diente. Die β-Glucosidaseaktivität wurde durch Inkubation (30 min bei 50°C) von 1 ml 0,029 M Cellobioselösung in 0,05 M Na-Acetatpuffer (pH = 4,6) mit 1 ml Enzymlösung bestimmt. Die Enzymkonzentration in dem Inkubationsansatz wurde so gewählt, daß eine lineare Abhängigkeit zwischen der Geschwindigkeit der Enzymreaktion und der Enzymkonzentration bestand. Die Messung der gebildeten Glucose erfolgte enzymatisch mittels Glucoseoxidase (Blutzucker-Testbesteck nach AB der DDR). Als eine Enzymeinheit wurde diejenige Menge an Enzym definiert, die unter Standardbedingungen 1 µM Glucose pro Minute bildet. Im verwendeten Cellulasekomplex betrug die spezifische β-Glucosidaseaktivität 0,80 U/mg Protein. Die Ermittlung kinetischer Konstanten der Cellobiosehydrolyse mittels linearer Regression wurde unter den Bedingungen der Aktivitätsbestimmung durchgeführt.

Ergebnisse und Diskussion

Die Auswertung der experimentellen Daten mit den linearisierten Geschwindigkeitsgleichungen ist in den Abbildungen 1 bis 3 enthalten. Die hierbei gewonnenen Ergebnisse unterscheiden sich nur geringfügig (Tabelle 1). Hinzuweisen ist auf die Verbreite-

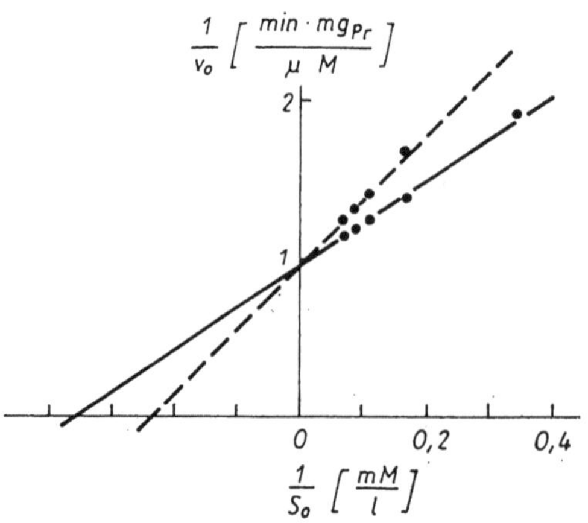

Abb. 1. LINEWEAVER-BURK-Diagramm unter Variation der Glucosekonzentration
(— 0 mM/m, --- 0,8 mM/ml)

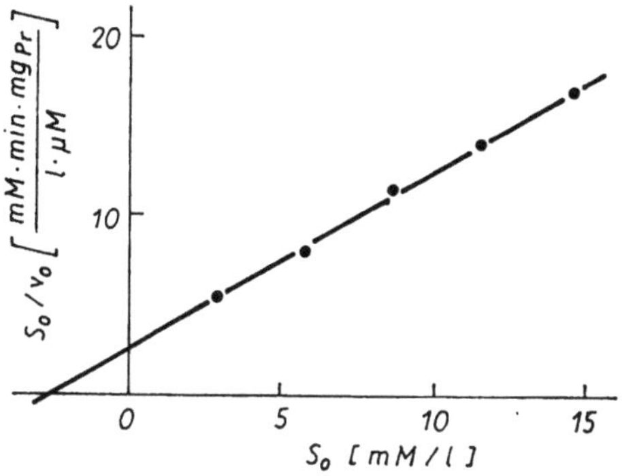

Abb. 2. Auftragung nach HANES

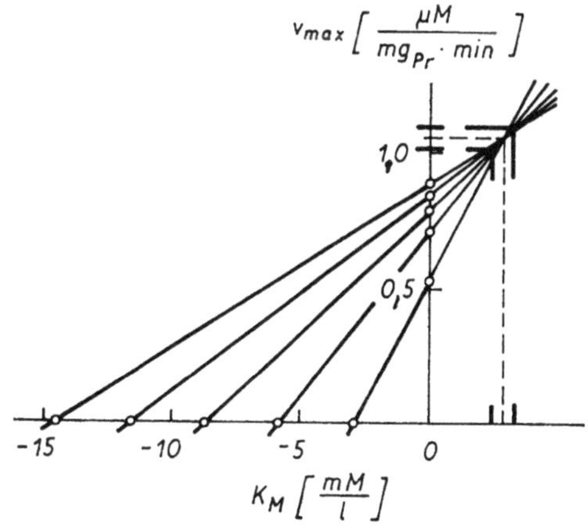

Abb. 3. Direkt-Plot-Methode zur v_{max}- und K_M-Bestimmung

rung des 95%-Konfidenzbereiches mit Zunahme der „statistischen Güte" des Auswerteverfahrens. Die Auftragung nach EISENTHAL/CORNISH-BOWDEN ergab nur dann befriedigende Geradenscharen, wenn die Inhibitoranfangskonzentration Null war. Durch die Variation der Anfangskonzentration an Glucose konnte der Wert von K_P mit 1,3 ± 0,2 mM/l aus dem LINEWEAVER-BURK-Diagramm ermittelt werden. Demnach handelt es sich beim Cellobioseabbau durch β-Glucosidase von *Gliocladium* sp. um eine kompetitive Hemmung durch das Produkt Glucose.

Für die nichtlineare Parameterbestimmung standen fünf Konzentration-Zeit-Kurven mit unterschiedlichen Anfangskonzentrationen an Cellobiose, Glucose und β-Gluco-

Tabelle 1. Kinetische Konstanten bei linearen Auswertemethoden

Parameter	LINEWAEVER-BURK	HANES	WILKINSON	EISENTHAL-CORNISH-BOWDEN
$k\left[\frac{\mu M_{Glucose}}{mg_{Protein} \cdot min}\right]$	1,04 ± 0,05	1,02 ± 0,09	1,06 ± 0,15	1,05 ± 0,03*
K_M [mM/l]	2,8 ± 0,3	2,6 ± 0,3	2,9 ± 0,9	2,9 ± 0,3*

* graphisch ermittelt

sidase zur Verfügung. Die Abbildung 4 zeigt eine der Versuchsreihen. Dabei ist die durchgezogene Kurve das Ergebnis der Parameterschätzung und die experimentell bestimmten Konzentrationswerte sind durch entsprechende Symbole dargestellt.
Bei der gemeinsamen Auswertung aller Konzentrationskurven ergab sich, daß neben dem auf Grund der Ergebnisse der linearen Regression erwarteten Fehlerquadratminimum ein weiteres Minimum bei ($k = 1{,}26$, $K_M = 6{,}9$, $K_P = 3{,}2$) in der Modellantwort-

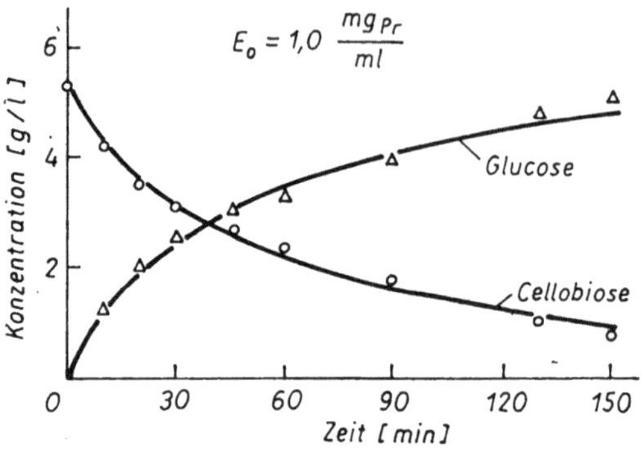

Abb. 4. Zeitlicher Verlauf der Cellobiosehydrolyse
(— Ergebnis der Parameterbestimmung)

fläche für den kompetitiven Hemmtyp existiert. Diese Doppeldeutigkeit der nichtlinearen Versuchsauswertung läßt sich durch weitere Versuche aufheben. Als mathematisches Detail ist zu erwähnen, daß sich das Determinantenkriterium von Box hinsichtlich der Konvergenzgeschwindigkeit der Schätzverfahren als günstigstes erwiesen hat.
Im Vergleich der nichtlinearen Versuchsauswertung zu den linearen Methoden fällt eine beachtliche Vergrößerung der Konfidenzbereiche auf — mit anderen Worten, die Ergebnisse sind noch „ungenauer". Dies trifft den tatsächlichen Sachverhalt aber nicht, denn die in der Lineweaver-Burk-Auftragung gefundenen Konstanten mit ihren Konfidenzbereichen können den in Abbildung 5 dargestellten zeitlichen Verlauf des Cellobioseabbaus nicht erklären. Zusammenfassend läßt sich für die Parameterschätzung feststellen

- lineare und nichtlineare Versuchsauswertung führen zu analogen kinetischen Konstanten
- alle linearen Auftragungen ergeben, bedingt durch die Hin- und Rücktransformation, zu schmale Konfidenzbereiche der Parameter.

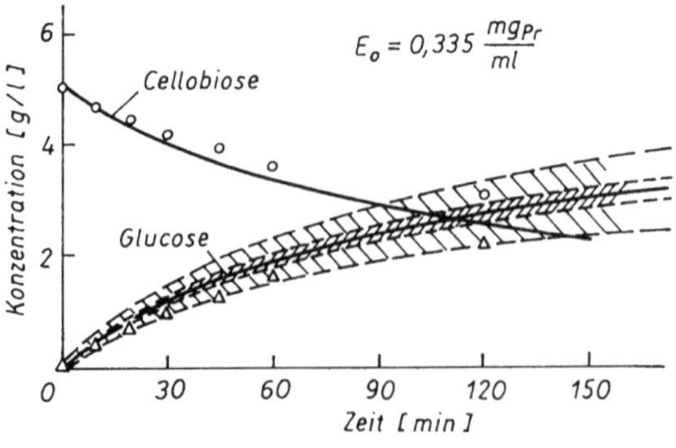

Abb. 5. Konfidenzband (95% Sicherheit) für lineare (/////) und nichtlineare (\\\\) Parameterschätzung

Von dieser bei den Methoden der linearisierten Geschwindigkeitsgleichungen vorgetäuschten Sicherheit wird auch die Unterscheidung der Hemmtypen betroffen. Wenn nicht mittels nichtlinearer Versuchsauswertung nachgewiesen werden kann, daß der aus den linearen Auftragungen erhaltene Hemmtyp eindeutig ist, muß die Entscheidung für diesen Hemmtyp fraglich bleiben. Von GONG et al. [11] wird eine nichtkompetitive Hemmung durch Glucose bei der Cellobiosehydrolyse vorgeschlagen. Dieser Arbeitsgruppe gelang die Auftrennung der Cellobiase von *Trichoderma viride* in drei Fraktionen. Die für zwei Fraktionen erhaltenen kinetischen Konstanten (Konfidenzbereiche wurden nicht ermittelt) sind in der Tabelle 2 enthalten. Im Vergleich zu den Ergebnissen der vorliegenden Arbeit ergeben die unterschiedlichen Proteingehalte von *Gliocladium* sp. und *Trichoderma viride* zwangsläufig verschiedene Geschwindigkeitskonstanten k, während für K_M und K_P eine zufriedenstellende Übereinstimmung gefunden wurde. Der Vergleich der Varianzen der Modelle über den F-Test zeigt eindeutige Vorteile der Modelle 2—5 gegenüber dem einfachen MICHAELIS-MENTEN-Ansatz. Die Unterscheidung zwischen kompetitiver (Modell 2) und nichtkompetitiver (Modell 3) Hemmung durch Glucose ist für zwei der fünf Konzentration-Zeit-Kurven nicht möglich. Die gemeinsame Auswertung aller Meßpunkte führt zu einer Überlegenheit des kompetitiven Hemmtyps, die erst mit zusätzlichen Versuchen signifikant wird [10]. Dem Modell 5 mit einer Enzyminaktivierung ist unter den gewählten Versuchsbedingungen dann der Vorzug gegenüber Modell 2 zu geben, wenn Versuche mit längeren Hydrolysezeiten (> 6 h) zur Auswertung gelangen.

Tabelle 2. Vergleich zwischen linearer und nichtlinearer Parameterschätzung

Parameter	lineare	nichtlineare	GONG [11]	
	Versuchsauswertung		Peak 2	Peak 3
$k \left[\frac{\mu M_{Glucose}}{mg_{Protein} \cdot min}\right]$	1,04 ± 0,05	1,02 ± 0,4	116	44,6
K_M [mM/l]	2,8 ± 0,3	2,6 ± 1,4	2,5	2,74
K_P (mM/l]	1,2 ± 0,2	1,2 ± 0,5	1,22	4,26

Eine sichere Entscheidung für einen wirklich zutreffenden Mechanismus liefert also bei der vorgegebenen Versuchsanzahl die nichtlineare Parameterschätzung nicht — sie macht jedoch auf diesen Umstand aufmerksam. Würde man sich auf den über lineare Auftragungen gewonnenen Hemmtyp beziehen, kann der in Abbildung 5 gemessene Kurvenverlauf zu einer widersprüchlichen Interpretation führen. So ließe sich der gegenüber der Theorie verringerte Glucosezuwachs u. a. durch einen zusätzlichen Entzug an freiem Enzym (Bildung eines Komplexes [ESP] — also nichtkompetitive Hemmung) erklären.

Bei einfachen Reaktionsmechanismen ist ein Vorteil der nichtlinearen Parameterschätzung insofern zu erkennen, da bei fehlerbehafteten Daten diese Unsicherheit sich im Auswerteergebnis wiederfindet und somit eine exakte Einschätzung der Untersuchung möglich wird und unter Umständen ein Fortsetzen der experimentellen Arbeiten anzuraten ist. Die Domäne der nichtlinearen Versuchsauswertung liegt jedoch auf dem Gebiet komplexer Reaktionen. Dies soll am Beispiel der enzymatischen Cellulosehydrolyse in einer weiteren Arbeit gezeigt werden.

Für die freundliche Unterstützung, die das Entstehen dieser Arbeit ermöglichte, möchten wir uns bei den Herren Prof. Dr. habil. H. KOCH und Dr. G. KLAPPACH herzlich bedanken.

Symbole

c_{ij}	gemessene Konzentration der Komponente i zur Zeit j
$f_i(t_j)$	errechnete Konzentration der Komponente i zur Zeit j
k, K_M, K_P	zu bestimmende Parameter der Modelle
n	Anzahl der Meßwerte
p	Anzahl der Modellparameter
t	Zeit
w_i	Gewichtsfaktor der Komponente i
$Z(k)$	Zielfunktion der Minimierung

Eingegangen: 10. 5. 1984

Literatur

[1] PÖHLAND, D., PRAUSE, M., ALBRECHT, Ch.: 2. Symposium der sozialistischen Länder über Biotechnologie. Leipzig, 1980.
[2] FROMMENT, G. F., MEZAKI, R.: Chem. Eng. Sci. **25** (1970), 293.
[3] EISENTHAL, R., CORNISH-BOWDEN, A.: Biochem. J. **139** (1974), 715.
[4] WILKINSON, G. N.: Biochem. J. **80** (1961), 324.
[5] MARQUARDT, D. W.: J. Soc. Ind. Appl. Math. **11** (1963), 431.
[6] FLETCHER, R., POWELL, M. J.: Computer J. **6** (1963), 163.
[7] BOX, G. E. P., DRAPER, N.: Biometrika **52** (1965), 355.
[8] VOTRUBA, J.: Acta Biotechnol. **2** (1982), 119.
[9] LOWRY, O. H., ROSEBOROUGH, N. J., FARR, A. L., RANDALL, R. J.: J. Biol. Chem. **193** (1951), 265.
[10] KUDE, J.: Dissertation (in Vorbereitung).
[11] GONG, C., LADISCH, M. R., TSAO, G. T.: Biotechnol. Bioeng. **19** (1977), 959.

Acta Biotechnol. 5 (1985) 3, 279–284

Метаболиты — индукторы автолиза дрожжей Saccharomyces cerevisiae

Бабаян Т. Л., Безруков М. Г. и Калунянц К. А.

Институт элементоорганических соединений им. А. Н. Несмеянова АН СССР
117813 г. Москва В-334
Вавилова 28, СССР

Summary

The accumulation of nucleic and protein (amino nitrogen) components as decomposition products of the process of autolysis of *Saccharomyces cerevisiae* yeast cells was studied at 50 °C under the effect of various membranotropic additives. The influence of the added n-alcohols, n-fatty acids and several peptides was investigated in the range of concentration of 0.1 – 0.5 M. The maximal acceleration of the autolysis has been demonstrated under the effect of additives with a hydrophobicity of 7.5 – 8.5 ccal/M. In all the investigated concentrations stearic acid and octadecyl alcohol have an inhibitory influence. The role of the hydrophobic influences and the mechanism of autolysis are discussed.

При развитии крупнотоннажных микробиологических производств необходимо учитывать воздействие компонентов культуральной среды на физиологическое состояние и химический состав дрожжевой биомассы в процессе ее непрерывного культивирования и дальнейшей переработки (сепарация, концентрирование, сушка и т. д.). Питательные свойства биомассы в большой мере зависят от ее физиологического состояния. Такие метаболиты, как аминокислоты, пептиды, углеводы, жирные кислоты, спирты и т.д., способные индуцировать автолитические процессы в клетках, влияют на молекулярно-весовые характеристики питательных компонентов и токсические свойства данной биомассы.

Как известно, автолиз представляет собой гидролитический распад внутриклеточных биополимеров под действием комплекса собственных ферментов клетки, причем индуцированным считается процесс, искусственно вызываемый посредством физических, химических или биологических агентов. В работе [1] показано, что такой автолиз является по крайней мере двухстадийным процессом: I — это перестройка белково-липидных эндоструктур, предшествующая активации эндоферментов и II — действие активированных гидролаз на соответствующие субстраты. Из продуктов метаболизма, присутствующих в культуральной жидкости при выращивании дрожжей, наиболее физиологически активными являются аминокислоты, пептиды, жирные кислоты и спирты. Роль аминокислот как индукторов автолиза была рассмотрена нами ранее [2]. В настоящей работе исследовалось воздействие н-спиртов, н-жирных кислот и некоторых пептидов на автолиз пекарских дрожжей.

Материалы и методы

Объектом исследования были хлебопекарные дрожжи, штамм 14 Л-1. Дрожжи не содержали посторонней микрофлоры, что было проверено путем высева дрожжевой биомассы на чашки Петри с МПА (мясо-пептонный агар) и сусло-агаром. Во всех опытах концентрация сухих веществ в дрожжевой суспензии составляла 20%.

Автолиз проводили в колбах, установленных на качалках в воздушном термостате при температуре 50° с точностью поддержания ее $\pm$ 1°. Пробы автолизатов отбирались из колб стерильно под пламенем. Динамику накопления продуктов автолиза в пробах автолизатов, подвергнутых центрифугированию, определяли по содержанию в надосадочной жидкости аминного азота [3] и нуклеиновых компонентов [4].

Навески соответствующих добавок были внесены в колбы с дрожжевой суспензией при комнатной температуре до помещения их в термостат. Вклад добавок по аминному азоту в надосадочной жидкости автолизата определяли экспериментально и учитывали при расчете эффективности автолиза.

Расчет проводили следующим образом: концентрацию аминного азота в контроле (без добавок) N_k за стандартное время 9 ч вычитали из концентрации аминного азота в опыте N_i. Эффективность автолиза (в процентах) рассчитывали отношением полученной величины к N_k, умноженным на 100. Каждая полученная цифра представляет среднюю арифметическую величину из трех независимых опытов. Точно так же рассчитывали эффективность автолиза по отношению к нуклеиновым компонентам клетки.

Активность α-химотрипсина в присутствии добавок исследовали модифицированным методом Ансона [5] на казеинате натрия в качестве субстрата, оба реактива производства НПО Биохимреактив г. Олайне, лецитин — фирмы Merck, ФРГ.

Результаты и их обсуждения

Исследуя механизм воздействия химических добавок на автолиз, необходимо учитывать следующее:

1) активацию протеолитических ферментов при освобождении их из лизосом;
2) воздействие этих ферментов на надмолекулярные липопротеиновые образования и расщепление биополимеров клетки;
3) взаимодействие продуктов расщепления биополимеров с ферментами (сопровождающееся или инактивацией или пролонгированием действия ферментов) и мембранами (сопровождающееся либо уплотнением, либо разрушением мембран).

Из литературы известно, что большая часть низкомолекулярных продуктов распада биополимеров клетки является ингибиторами литических ферментов [6]. Нами ранее было показано, что в некоторых случаях ингибирование аминокислотами процесса автолиза не наблюдалось для наиболее гидрофобных аминокислот, что может быть объяснено их солюбилизацией липидными компонентами клетки, приводящей к падению их активной концентрации [2]. Воздействие гидрофобных добавок на прочность внутриклеточных структур достаточно интересно, как с чисто практической точки зрения, так и для выяснения механизма контактов между биополимерами в органеллах клетки.

Поскольку аминокислоты ограничивают область гидрофобности 3000 кал/М, мы продолжили изучение процесса автолиза *S. cerevisiae* в присутствии алифатических спиртов, гидрофобность которых прямо пропорциональна количеству углеродных атомов в цепи и позволяет значительно расширить диапазон гидрофобности исследуемой добавки. Любая добавка проявляет свое активирующее действие на автолиз лишь в определенной концентрации. Так из рис. 1 видно, что ускоряющее воздействие этилового спирта на автолиз имеет колоколообразный профиль с

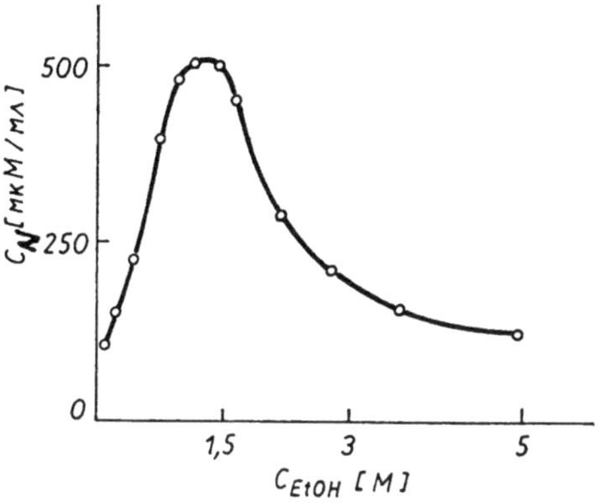

Рис. 1 Зависимость концентрации аминного азота в автолизате за 9 ч автолиза хлебопекарных дрожжей от концентрации добавленного EtOH.

оптимумом 1,3 М. Оптимальная концентрация имеет тенденцию уменьшаться с ростом гидрофобности добавки от 1,3 М (для EtOH) к 0,2 М (для н-додецилового спирта). Для наиболее активного н-нонилового спирта (табл. 1), который обеспечивает ускорение автолиза более чем в 3 раза по сравнению с контролем, оптимум составляет 0,4 М.

Такой же характер носит активирующее воздействие кислот (табл. 2), среди которых наиболее активна лауриновая (с HP = 7,8 ккал/моль), ускоряющая автолиз почти в 4 раза.

Таблица 1. Оптимальная концентрация добавки, обеспечивающая максимальное ускорение автолиза

Добавки	Гидрофобность HP [ккал/M]	Концентрация C [M]	Эффективность автолиза в % по отношению к контролю
Контроль (дрожжевая суспензия без добавок)	0	0	100
н-спирты:			
этиловый	1,630	0,90	220
пропиловый	2,445	0,45	205
бутиловый	2,360	0,35	200
гексиловый	5,190	0,35	220
октиловый	6,520	0,30	235
нониловый	7,335	0,40	315
додециловый	9,780	0,20	240
октадециловый	14,670	0,05	40
н-жирные кислоты			
уксусная	0,710	0,10	270
пропионовая	1,420	0,10	250
каприловая	4,970	0,10	340
лауриновая	7,800	0,30	380
миристиновая	9,230	0,30	350
стеариновая	12,070	0,30	50

Таблица 2. Эффективность автолиза по отношению к контролю при концентрации добавок 0,3 М

Добавка	Гидрофобность HP [ккал/М]	выход NH_2-групп в %	выход нуклеиновых компонентов в %
н-спирты:			
этиловый	1,630	90	100
пропиловый	2,445	120	130
бутиловый	3,260	200	220
гексиловый	5,190	220	230
октиловый	6,520	230	250
нониловый	7,335	310	340
додециловый	9,780	230	225
октадециловый	14,670	0	0
н-жирные кислоты:			
уксусная	0,710	240	260
пропионовая	1,420	240	270
каприловая	4,970	290	300
лауриновая	7,800	380	400
миристиновая	9,230	350	370
стеариновая	12,070	50	50
пептиды:			
аланил-серин	0,710	125	130
аналин-аналин	1,500	80	80
аланил-метионин	2,050	75	75
аланил-валин	2,450	70	70
аланил-лейцин	3,150	50	60
тетрапептид:* Z-трип-ала-гли--гли-OH	6,35	145	160
гексапептид: тир-гли-лей--арг-прол-глицин	8,60	320	300

*Карббензоксипроизводное этого пептида

На рис. 2 приведены данные по ускорению автолиза эквимолярными количествами (0,3 М) спиртов, кислот и пептидов в зависимости от гидрофобности добавки. Обращает на себя внимание максимальное увеличение активности всех классов добавок, имеющих гидрофобность 7,5—8,5 ккал/моль. При увеличении гидрофобности добавок выше 8,5 ккал/моль их активность падает. Пептиды, обладающее большей гидрофобностью, не были исследованы из-за недоступности чистых реактивов. Следует отметить, что у C_{18} спирта и кислоты был выявлен только тормозящий эффект во всем изученном интервале концентрации.

По-видимому, можно предположить, что в активации автолитических процессов в клетке существенную роль играет гидрофобно-гидрофильний баланс (дифильность) индуктора и его взаимодействие с липопротеиновыми структурами клетки. Об этом косвенно свидетельствует аналогия в активации как протеолитических ферментов, так и нуклеаз. Воздействие изученных добавок (кроме пептидов) на свободные ферменты можно считать менее существенным. Хотя в литературе имеются противоречивые и взаимоисключающие сведения о воздействии дифильных

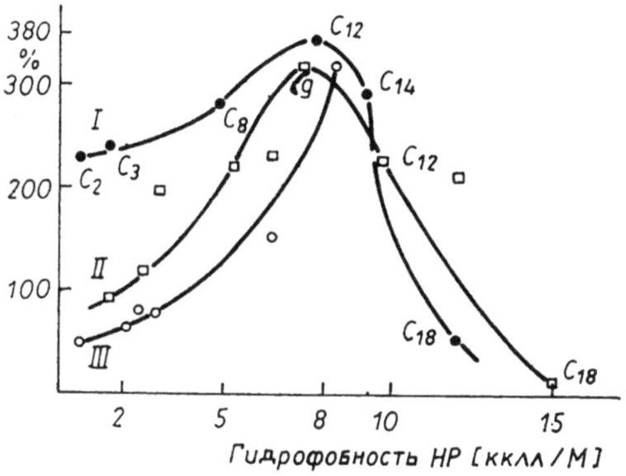

Рис. 2 Воздействие различных добавок на эффективность автолиза в зависимости от их гидрофобности (концентрация всех добавок 0,3 М); I ● — кислоты, II □ — спирты, III ○ пептиды.

молекул на ферменты [7, 8], наши данные (табл. 3) на примере химотрипсина свидетельствуют о несущественном воздействии подобных молекул на свободные ферменты, осуществляющие собственно гидролиз биополимеров в процессе автолиза. Характерно, что даже лецитин, являющийся поверхностноактивным агентом, и, одновременно, компонентом липопротеиновых эндоструктур клеток [9], не оказывает заметного воздействия на уровне свободного фермента.

Результаты, полученные в данной работе, свидетельствуют о том, как важно учитывать состав культуральной жидкости в процессах выращивания и переработки биомассы микроорганизмов. Поскольку культуральная жидкость содержит гораздо большее количество компонентов, чем изученное в настоящем исследовании, то их реальное воздействие на автолитические процессы в биомассе может быть гораздо более существенным. Из литературы, например, известно о синергизме таких активаторов автолиза, как этанол и NaCI [10], этилацетат и органические кислоты [11].

В культуральной жидкости присутствуют как моно-, так и поливалентные катионы, которые, безусловно, вступят в синергизм с изученными метаболитами. Следовательно, в каждом конкретном случае для каждого культивируемого штамма микроорганизма необходимо экспериментальное изучение воздействия реальной культуральной жидкости на физиологическое состояние культуры.

Таблица 3. Воздействие химических агентов на активность химотрипсина

Добавки		Активность α-химотрипсина в % по отношению к контролю
Контроль: без добавок (субстрат — казеинат натрия)		100
	концентрация	
EtOH	2%	90
	4%	110
	6%	95
лецитин	0,002 г/мл	90
	0,010 г/мл	110
лауриновая к-та	0,1 м	90
	0,05 м	110

Выводы

1. Изученные добавки обладают концентрационным оптимумом индуцирования автолиза. Эта концентрационная зависимость имеет колоколообразный профиль. Октадециловый спирт и стеариновая кислота оказывают тормозящий эффект во всем изученном интервале концентраций.
2. Наибольшую активность проявляли во всех исследованных классах соединения, характеризующиеся гидрофобностью 7,5—8,5 ккал/М.
3. Основная функция изученных добавок по всей вероятности заключается в разрушении надмолекулярных внутриклеточных структур, т.е. в ускорении первой стадии процесса автолиза.

Авторы принося свою благодарность В. Т. Иванову за предоставленный им гексапептид, синтезированный в его лаборатории, и за ценные рекомендации при обсуждении данной работы.

Поступила в редакцию: 15. 12. 1983

Литература

[1] Babayan T. L., Bezrukov M. G., Latov V. K., Belikov V. M., Belavtseva E. M., Titova E. F.: Current Microbiology **5** (1981), 163.
[2] Babayan T. L., Bezrukov M. G., Latov V. K., Belikov V. M.: Current Microbiology, **2** (1979), 301.
[3] Fields, R.: Biochem. J., 124 (1971), 581.
[4] Спирин А. С.: Биохимия **23** (1958), 656.
[5] Васильева Р. П., Евтихов П. Н., Богорад Г. В.: Микробиолог. промышленность № 4 (1976), 23.
[6] Мартинкова Н. С., Витт С. В., Беликов В. М.: Микробиолог. промышленность **6** (1971).
[7] Мартинек К., Левашов А. В., Березин И. В.: Молекулярная Биология **4** (1970) 517.
[8] Мосолов В. В., Афанасьев П. В.: ДАН СССР, **152** (1963) № 3, 748.
[9] Покровский А. А., Тутельян В. А.: Лизосомы, Москва, Наука, 1977.
[10] Hiroshi, Sugimoto: J. Food Sci. **39** (1974), 5, 939.
[11] Авт. свид. СССР № 552953, 1976, БИ, 1977а, № 13, 13.

Acta Biotechnol. 5 (1985) 3, 285—291

Wirkung von Ölen und Fettsäuren auf Wachstum und Enzymbildung bei Thermoactinomyces vulgaris

IV. Einfluß von Art und Menge des Öles sowie des Zeitpunktes der Ölzugabe

LEUCHTENBERGER, A. und RUTTLOFF, H.

Akademie der Wissenschaften der DDR
Forschungszentrum für Molekularbiologie und Medizin
Zentralinstitut für Ernährung
DDR-1505 Potsdam-Rehbrücke, Arthur-Scheunert-Allee 114—116

Summary

The effect of different quantities and kinds of vegetable and animal oils and the importance of the time of the oil addition on growth and protease synthesis by *T. vulgaris* was investigated. The used oils stimulate the protease production if they are added in a suitable concentration. However, the stimulation effect of each oil is different. Culture inoculation with spores gives the best results, when oil was added to the medium 1 up to 2 hours after beginning of the fermentation. The enzyme activity is equal to or lower than the control, when the oil addition was carried out before or 3 hours after starting the fermentation.

Einleitung

Aus dem in Mitteilung I zusammengestellten Literaturüberblick [1] geht u. a. hervor, daß die Wirkung von Lipiden auf mikrobielle Stoffwechselprozesse häufig von Art und Menge des verwendeten Öles bzw. der eingesetzten Fettsäuren sowie vom Zeitpunkt der Lipidzudosierung abhängig ist. So berichten z. B. EL SAID und GHALI [2], daß die Bildung von Protease und Amylase durch *Bacillus subtilis* von 1 bis 2% Reiskleieöl in einem komplexen Nährmedium um 20 bis 30% erhöht wird, wohingegen Leinsaat-, Sesam-, Mais-, Baumwollsaat- und Walöl in gleicher Konzentration nicht oder sogar hemmend wirken. Nach KOBAYASHI u. SUZUKI [3] wird die α-Galaktosidasebildung bei *Mortierella vinacea* durch Zugabe von 0,1 bis 0,3% C_1- bis C_{11}-Fettsäuren gehemmt, während C_{12}- bis C_{20}-Fettsäuren diese aktivieren. Gemäß KONDO [4, 5] ist die Proteasewirkung von *Bacillus subtilis* und *Streptomyces griseus* auf verschiedene Proteine von der Konzentration der zugegebenen C_8- bis C_{12}-Fettsäuren abhängig. Unterschiedliche Auswirkungen auf die Proteasebildung in Abhängigkeit vom Zeitpunkt der Ölzugabe zum Fermentationsmedium beobachten ARAVINA et al. [6] sowie GALYNKIN et al. [7]. Die folgenden Untersuchungen beschäftigen sich mit dem Einfluß dieser Parameter auf Wachstum und Proteasebildung bei *T. vulgaris*.

Material und Methoden

Der Stamm *T. vulgaris* wird bis zu 4 Wochen auf Maisquellwasser (MQW)I-Agar bei 4°C längerfristig als Sporenkonserve in Luvos-Heilerde [8, 1] gelagert. Die Zusammensetzung der verwendeten Medien (MQW I-Agar, MYP-Medium, MQW I-, MQW II- und MQW V-Medium), die Durchführung der Fermentation in 500 ml-Einstich(E)-Kolben, in 2 l- und 30 l-Rührfermentoren sowie das Ansetzen von Vor- und Arbeitskultur sind in Mitteilung III [9] näher ausgeführt. Die Wachstumsmessung erfolgt durch Proteinbestimmung im Sediment von 3 ml Kulturmedium [10]. Die Proteaseaktivität wird nach einer modifizierten ANSON-Methode unter Verwendung einer 1%igen Caseinlösung, nach Inkubation bei 55 °C über 20 min [8] bestimmt. Die dargestellten Ergebnisse sind Mittelwerte von mindestens 3 Versuchsansätzen.

Ergebnisse

Wirkung verschiedener pflanzlicher und tierischer Öle

In Versuchsansätzen mit 500 ml-Einstichkolben wird der Einfluß der qualitativen Zusammensetzung verschiedener Öle auf die Proteasebildung von *T. vulgaris* untersucht. Die in MQW I-Medium bei Zusatz von 0,1% Raps-, Soja-, Fisch- und Rinderklauenöl erzielten Ergebnisse sind in Abb. 1 enthalten.

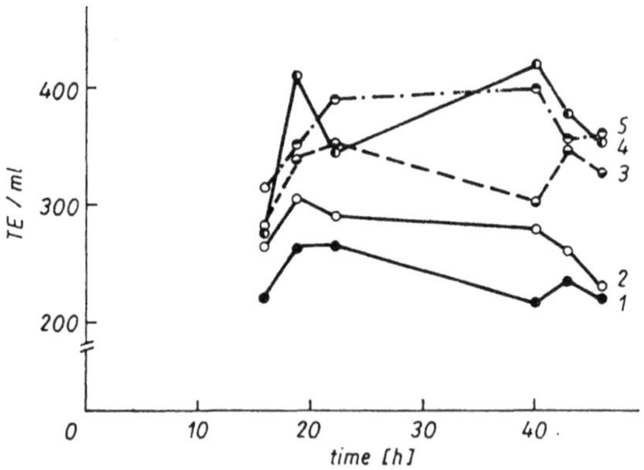

Abb. 1. Beeinflussung der Proteasebildung in 500 ml-Einstichkolben durch Zusätze von jeweils 0,1% verschiedener Öle (MQW I-Medium, 46 h Kultivierung)

1 Kontrolle
2 Rinderklauenöl
3 Rapsöl
4 Fischöl
5 Sojaöl

Sämtliche Öle bewirken eine erhöhte Proteaseausschüttung; der stimulierende Effekt ist jedoch unterschiedlich. Nach 19 bis 22 h Fermentationsdauer ergeben Zusätze von Rinderklauenöl etwa um 25%, von Rapsöl um 40%, von Fisch- und Sojaöl um 50 bis 60% höhere Enzymausbeuten als der Kontrollansatz.

Eine umfassende Testung unter Verwendung von 10 verschiedenen tierischen und pflanzlichen Ölen bzw. Fetten wird in dem nährstoffreicheren MQW II-Medium durchgeführt (Abb. 2). Nach 16 bis 19 h Fermentationsdauer zeigen auch in diesem Medium alle Öle eine aktivitätssteigernde Wirkung. Der absolute Zuwachs ist geringer als in MQW I-Medium. Die Aktivitätserhöhung durch die einzelnen Öle ist unterschiedlich und liegt zwischen 20 und 60% im Vergleich zur Kontrolle. Die beste Wirkung wird auch in diesem Medium mit Fischöl erzielt.

Variation der Rapsölmenge

Untersuchungen im 30 l-Fermentor haben bereits ergeben, daß zwischen der zudosierten Ölmenge und der Proteasebildung ein Zusammenhang besteht [1]. Versuche mit 0,1 bis 0,8% Rapsölzusatz in 500 ml-Einstichkolben bestätigen diesen Befund (Tab. 1). Die optimale Konzentration liegt hier bei 0,1% Öl.

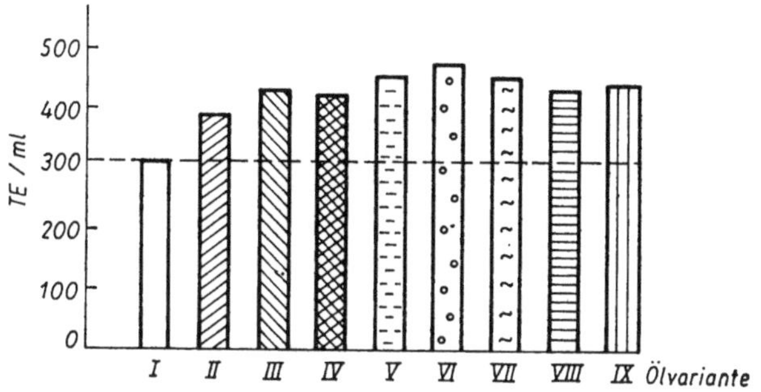

Abb. 2. Einfluß verschiedener tierischer und pflanzlicher Öle auf die Proteasebildung in 500 ml-Einstichkolben (MQW II-Medium, 19 h Kultivierung)

I Kontrolle
II Rapsöl
III Maiskeimöl
IV Rüböl
V Kokosraffinat
VI Fischöl
VII Rohleberöl (Fischöl)
VIII Sojaöl
IX Sonnenblumenöl

Tabelle 1. Einfluß der Rapsölkonzentration auf die Proteasebildung in 500 ml-Einstichkolben (MQW I-Medium, 50 °C, 18 h Kultivierung)

Rapsölmenge %	0	0,1	0,2	0,4	0,8
Proteaseaktivität TE/ml	210	420	360	210	222

Zeitpunkt der Ölzugabe

Von Einfluß ist auch der Zeitpunkt des Ölzusatzes. Versuche in MQW V-Medium liefern bei Zugabe von 0,1% Rapsöl in 500 ml-Einstichkolben die in Abb. 3 enthaltenen Ergebnisse. Danach werden in Schüttelkultur die besten Resultate erzielt, wenn das Öl 1 bis 2 Stunden nach Fermentationsbeginn (Anfang der log-Phase) zum Kulturmedium gegeben wird. Wird das Öl hingegen vor Fermentationsbeginn bzw. erst 3 h danach

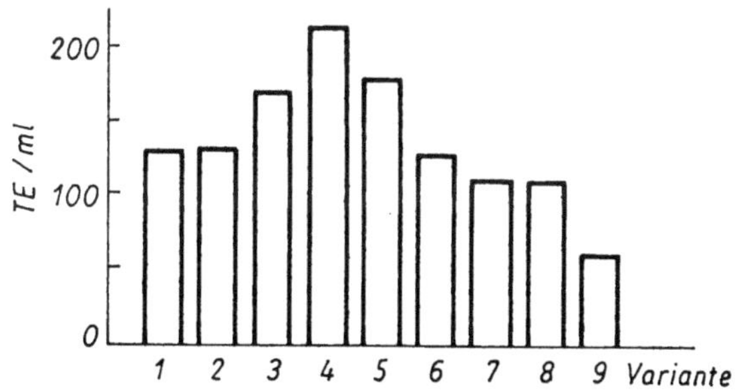

Abb. 3. Beziehung zwischen dem Zeitpunkt der Ölzugabe und der Proteasebildung in 500 ml-Einstichkolben (MQW V-Medium, 0,1% Rapsöl, 24 h Kultivierung)

1 — Kontrolle, 2 — vor Mediensterilisation, 3 — t_0, 4 — t_1, 5 — t_2, 6 — t_3, 7 — t_4, 8 — t_5, 9 — t_6.

und später dem Kulturmedium zugefügt, ist die Enzymausbeute gleich bzw. sogar geringer als bei der Kontrolle. Daraus wird abgeleitet, daß Ölzugaben vor bzw. zu Beginn der Fermentation eine Verzögerung der Sporenkeimung und des Wachstums herbeiführen.

Untersuchungen hierzu bestätigen diese Annahme. Die Ermittlung der Keimungsrate erfolgt durch Auszählung von etwa 300 bis 400 Sporen und des davon ausgekeimten Anteils in mehreren mikroskopischen Blickfeldern. Der Mittelwert mehrerer Versuchsansätze ist in Abb. 4 dargestellt. Daraus ist zu ersehen, daß unter Schüttelkulturbe-

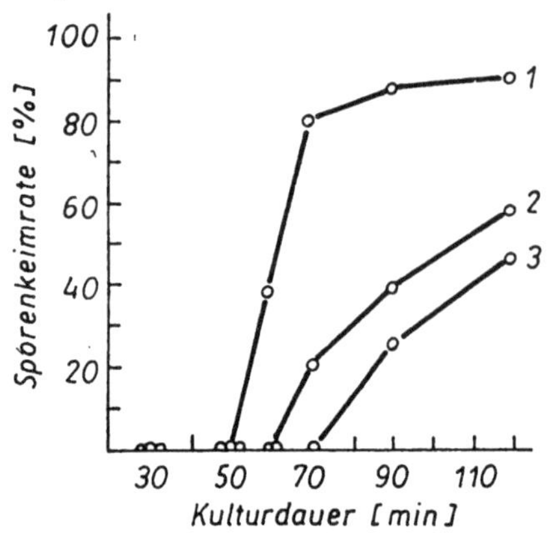

Abb. 4. Beeinflussung der Sporenkeimung durch Zugabe von Öl in 500 ml-Einstichkolben (MQW V-Medium, 0,1% Rapsöl).
1 — ohne Ölzusatz
2 — Ölzusatz nach Mediensterilisation
3 — Ölzusatz vor Mediensterilisation.

dingungen in Einstichkolben die Sporenkeimung in der Kontrolle nach 50 min einsetzt und innerhalb von 20 min 80% der Sporen ausgekeimt vorliegen; danach verläuft der Keimungsprozeß langsamer. Die Keimung in Gegenwart von Öl beginnt generell später und erreicht bis zur 120. Minute nur eine Keimungsrate von etwa 50%. Von Bedeutung ist noch die Beobachtung, daß die Zugabe von Öl vor der Mediensterilisation für die Auskeimung der Sporen von besonderem Nachteil ist. Die unterschiedliche Keimungsrate zum Zeitpunkt 70 min nach Kulturbeginn mit und ohne Ölzusatz ist aus Abb. 5 zu entnehmen.

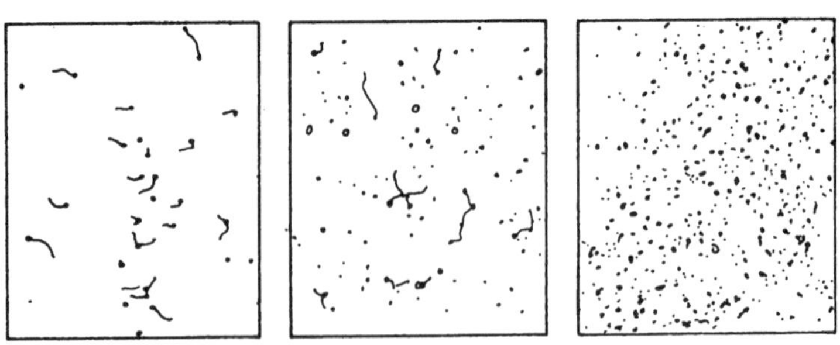

Bild a Bild b Bild c

Abb. 5. Keimungsrate von Sporen bei An- bzw. Abwesenheit von Öl 70 min nach Fermentationsbeginn in 500 ml-Einstichkolben (MQW V-Medium, 0,1% Rapsölzusatz). Vergrößerung 1 : 1000.

Bild a: ohne Ölzusatz
Bild b: Ölzusatz nach Mediensterilisation
Bild c: Ölzusatz vor Mediensterilisation.

Die Auswirkungen des Zeitpunktes der Ölzugabe zum Kulturmedium auf das Wachstum sind aus Abb. 6 ersichtlich. Danach führt Öl — unabhängig vom Zeitpunkt der Zugabe — im Vergleich zur Kontrolle zu einem längeranhaltenden Biomassewachstum. Während in der Kontrolle bereits nach 4 h das Maximum der Biomassebildung vorliegt, ist dieses bei den Ölvarianten erst nach 8 h erreicht. Die nach diesem Zeitraum gebildete Biomassemenge liegt um etwa 80% höher als die maximale Biomassemenge im Kontrollansatz nach 4 h. Unterschiede — in Abhängigkeit vom Zeitpunkt der Ölzugabe — zeigen sich in der Anfangsphase des Wachstums. Wird das Öl vor der Sterilisation des Mediums zugesetzt, beginnt die Biomasseentwicklung etwa eine Stunde später. Etwas verlangsamt ist das Wachstum auch bei der Zugabe des Öles zum Zeitpunkt t_1, während eine Ölzuführung zu einem späteren Zeitpunkt keine Unterschiede zur Kontrolle bewirkt.

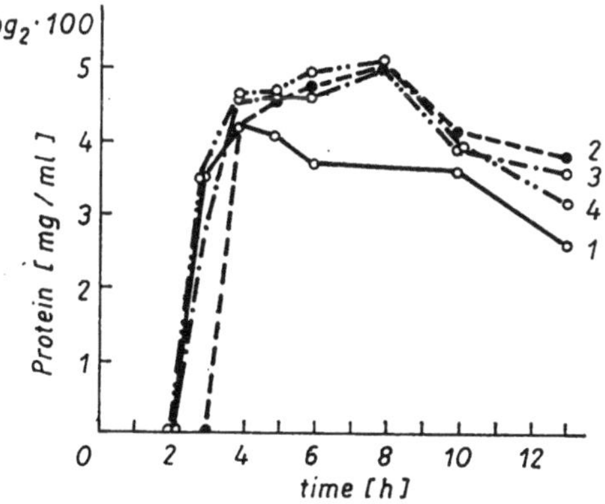

Abb. 6. Einfluß des Zeitpunktes der Ölzugabe auf das Wachstum in 500 ml-Einstichkolben (MQW V-Medium, 0,1% Rapsöl).

1 — ohne Ölzusatz
2 — Ölzusatz vor Mediensterilisation
3 — Ölzusatz t_1
4 — Ölzusatz t_6

Tabelle 2. Statistische Auswertung der Ergebnisse von Schüttelkulturansätzen mit und ohne Rapsölzusatz (MQW II-Medium, 20 h, 0,1% Rapsöl)

Erläuterungen: K — Kontrolle, Öl — Variante mit Ölzusatz

Versuchsansatz	Variante	n	$\bar{x}$ TE/ml	$\bar{x}$ %	sx TE/ml	s$\bar{x}$ TE/ml	V %	t-Test K → Öl
I	K	4	257,2	100	13,0	6,5	5,0	0,05%
	Öl	4	383,5	149	35,4	17,7	9,2	
II	K	9	286,7	100	64,0	21,3	22,3	0,05%
	Öl	9	486,7	170	88,7	29,5	18,2	
III	K	12	371,2	100	70,3	20,3	18,9	0,05%
	Öl	12	693,7	187	122,3	35,3	17,6	
⌀	K	9	371,5	100	170,4	56,8	45,8	0,5%
I–III	Öl	9	564,1	152	46,2	15,4	8,1	

Signifikanznachweis des Öleffektes in Schüttelkultur

Zur statistischen Absicherung der bei Zusatz von Öl beobachteten erhöhten Proteaseausschüttung werden mehrere Versuchsansätze ausgewertet. Unter Berücksichtigung der Wirkung verschiedener Einflußfaktoren werden 3 Stammkonserven unterschiedlicher Leistung, MQW II-Medium sowie 500 ml-Rundkolben mit $4 \times 0{,}5$ cm tiefen Schikanen verwendet. Die Zugabe von 0,1% Rapsöl erfolgt zu Fermentationsbeginn. Nach 20 h Kulturdauer werden Resultate gemäß Tab. 2 erhalten. Die Ergebnisse weisen aus, daß die Proteaseaktivität der Ölvariante signifikant höher liegt als bei der Kontrolle. Die Aktivitätserhöhung durch Ölzugabe beträgt in diesem Medium 50 bis 85%.

Diskussion

Übereinstimmend mit Befunden anderer Autoren wird der „Öleffekt" sowohl von der Konzentration als auch der Art des Öles beeinflußt. In Abhängigkeit vom verwendeten Stamm und den Kulturbedingungen gibt es mehr oder weniger große Unterschiede hinsichtlich der optimalen Mengen und der Ölart (Abb. 1). Der bei *T. vulgaris* ermittelte optimale Konzentrationsbereich von 0,1 bis 0,3% Rapsöl stimmt mit Ergebnissen von EKLUND et al. [11], ARAVINA et al. [6] sowie YOKOTSUKA et al. [12] überein. Gemäß Abb. 1 und 2 zeigen hierbei Fischöl, Sojaöl, Sonnenblumenöl, Rapsöl sowie Maiskeimöl die größte Wirkung; diese Substrate werden auch von YAMAMOTO et al. [13], RUTTLOFF et al. [14], YOKOTSUKA et al. [12], EKLUND et al. [11], GALYNKIN et al. [7] sowie PACA [15] als diesbezüglich besonders vorteilhaft angeführt. Hinsichtlich der Rangfolge ihrer Effektivität finden sich allerdings — je nach verwendetem Enzymbildner — bei den genannten Autoren unterschiedliche Angaben. Anders als bei der mit *T. vulgaris* gefundenen Reihenfolge Fischöl, Sonnenblumenöl, Sojaöl, Rapsöl halten z. B. YAMAMOTO et al. [13] Rapsöl für den effektivsten Stimulator, dann folgen Sojaöl, Fischöl und andere. Nach FRANZKE et al. [16] ist Maiskeimöl wirksamer als Rapsöl. In Abhängigkeit von den vorherrschenden Kulturbedingungen gibt es Unterschiede in der Metabolisierung der Entschäumeröle bzw. ihrer Fettsäuren durch verschiedene Mikroorganismen, die stammspezifisch und selten verallgemeinerungsfähig sind.

Der Zeitpunkt der Ölzugabe zum Fermentationsmedium zeigt gemäß Abb. 3 eine erhebliche Auswirkung auf die Proteasebildung, wenn mit Sporen beimpft wird. Die höchste Enzymaktivität wird erhalten, wenn Rapsöl 1 bis 2 h nach Fermentationsbeginn zugesetzt wird. Frühere wie auch spätere Zudosierungen führen zu merklichem Aktivitätsabfall nach 24 h Kulturdauer. Über ähnliche Verhältnisse berichten ARAVINA et al. [6], die bei Zugabe von 0,1% Sonnenblumenöl zu Beginn der Fermentation sogar einen Rückgang der Proteasebildung durch *Aspergillus terricola* um 14% gegenüber der Kontrolle feststellen. Bei Zugabe des Öles nach 24 h ermitteln sie nach 3tägiger Kultur eine Enzymaktivität, die um ca. 10% über dem Kontrollwert liegt. GALYNKIN et al. [7] finden bei *Actinomyces hygroscopicus*, daß bei Zugabe von 1% Zahnwalöl 2 Tage nach Fermentationsbeginn und später nur noch 10% der bei früherer Ölzugabe gemessenen Aktivität vorliegen.

Zur Ermittlung der Ursache wird der Einfluß des Ölzusatzes auf die Sporenkeimung sowie auf das Mycelwachstum geprüft. Gemäß Abb. 4 wirken sich Ölzusätze zu Beginn der Fermentation hemmend auf die Sporenkeimung aus, was im Endeffekt zu einer verzögerten Mycelentwicklung führt. Die stärkere Behinderung der Sporenkeimung bei Ölzusätzen vor der Sterilisation dürfte mit der beim Erhitzungsprozeß erfolgten intensiveren Ölverteilung bzw. mit der bakteriziden Wirkung freigesetzter Fettsäuren zu erklären sein. Die Hemmung der Sporenkeimung hat auch ein verzögertes Wachstum zur Folge (Abb. 6), welches jedoch im Vergleich zu den anderen Ölvarianten allmählich

ausgeglichen wird. Nach 8 h Kulturdauer gibt es keinen Unterschied in der Biomassemenge der Ölvarianten; sie ist doppelt so hoch wie bei der Kontrolle. Daraus wird geschlußfolgert, daß die durch verschiedene Zugabezeiten des Öles verursachte Aktivitätsunterschiede nicht wachstumsbedingt, sondern die Folge einer Beeinflussung der Enzymsynthese sind. Diese könnte direkt oder indirekt über die Veränderung des Stoffaustausches durch die Zellmembran erfolgen.
Die statistische Auswertung von Versuchsansätzen in 500 ml-Einstichkolben unter Berücksichtigung der als vorteilhaft ermittelten Einflußfaktoren belegt, daß die proteolytische Aktivität durch Rapsöl auch in Schüttelkultur signifikant um ca. 70% gesteigert wird (Tab. 2).

Für ihre gewissenhafte experimentelle Arbeit danken wir Frau Ursula Behrens und Frau Ursula Müller.

Eingegangen: 26. 3. 1984

Literatur

[1] LEUCHTENBERGER, A., RUTTLOFF, H.: Z. Allg. Mikrobiol. **19** (1979), 609.
[2] EL SAID, H., GHALI, Y.: Process Biochem. **11** (1976), 42—43, 45.
[3] KOBOYASHI, H., SUZUKI, H.: J. Ferment. Technol. **50** (1972), 835.
[4] KONDO, M.: J. Biochem. **45** (1958), 151, zitiert bei KONDO [17].
[5] KONDO, M.: J. Biochem. **46** (1958), 705, zitiert bei KONDO [17].
[6] ARAVINA, L. A., PONOMAREVA, V. D., TERESHIN, J. M., KASATKINA, J. D., GREKOVA, V. K.: Mikrobiologija (Moskva) **45** (1976), 770.
[7] GALYNKIN, V. A., EFIMOVA, T. P., GREKOVA, V. K., KOROLEVA, T. A.: Khim.-Farm. Zh. **12** (1978), 116.
[8] KLINGENBERG, P., ZICKLER, F., LEUCHTENBERGER, A., RUTTLOFF, H.: Z. Allg. Mikrobiol. **19** (1979), 17.
[9] LEUCHTENBERGER, A., RUTTLOFF, H.: Z. Allg. Mikrobiol. **23** (1983), 635.
[10] LEUCHTENBERGER, A., SCHAFFNER, B., RUTTLOFF, H.: Z. Allg. Mikrobiol. **21** (1981), 671.
[11] EKLUND, E., HEINONEN, H., GYLLENBERG, H. G.: Zbl. Bakteriol. Abt. II., **129** (1974), 12.
[12] YOKOTSUKA, T., HASHIMOTO, H., IWAASA, T.: US-Patent 367 4644.
[13] YAMAMOTO, T., HATTORI, F., TAKATSU, H.: Bull. Inst. chem. Res. Kyoto **42** (1964), 252.
[14] RUTTLOFF, H., TÄUFEL, A., ZICKLER, F., SCHIERBAUM, F., RICHTER, M., KUPKE, G., HUBER, J., RIEDEL, K.: DD-WP 111 525.
[15] PACA, J.: Kvasny Prumysl **24** (1978), 86.
[16] FRANZKE, C., GÖBEL, R., KROLL, J., LIEBS, P., RITTER, G., SCHULTZE, M.: Nahrung **23** (1979), 283.
[17] KONDO, M.: J. Biochem. **52** (1962), 279.

Acta Biotechnol. 5 (1985) 3, 292

Book Review

REHM, H.-J. (Herausgeber)

Biotechnologie 83

Mikrobielle Stoffproduktion, Neuentwicklungen in der Biotechnologie, Gentechnik

Vorträge von der Jahrestagung 1983 der Biotechnologen,
Dechema-Monographien, Band 95

Weinheim/Deerfield Beach, Florida/Basel: Verlag Chemie, 1984.
341 S., 156 Abb., 49 Tab., 281 Lit., 30 DM.

Die drei Abschnitte des Buches sind durch je 9 Vorträge repräsentiert. Bei den mikrobiellen Produkten liegt der Schwerpunkt bei den Antibiotika und Enzymen. Je ein Beitrag zur Glycerinbildung und Ethanolfermentation führt in den Bereich der Chemikalien und Grundstoffe. Für die biotechnischen Neuentwicklungen sind Mitteilungen über die Gestaltung von Bioreaktoren und die dabei vorherrschenden Tendenzen, wie Berechnung und Abschätzung thermodynamischer Daten, Reaktions- und Reaktormodellierung, Regelung und Erarbeitung optimaler Prozeßstrategien ebenso kennzeichnend wie Verfahren zur Aufarbeitung gentechnisch erzeugter Produkte. Im Abschnitt über Gentechnik werden anschauliche Beispiele für die Perspektive dieser Methoden in Industrie und Landwirtschaft gegeben. Stofflich gesehen betreffen sie die Produktion von Saccharose, Ethanol und die N_2-Fixierung. Unter Aspekten der multivalenten methodischen und konzeptionellen Entwicklungen werden die chemische DNA-Synthese (WINNACKER), die Genmanipulation mit mitochondrialer DNA bei höheren Organismen (ESSER), Klonierungssysteme bei Hyphenpilzen (TUDZYNSKI) und die Temperaturstreß-regulierte Genexpression bei höheren Pflanzen (SCHÖFFL) behandelt. Die Beiträge sind von unterschiedlichem Gewicht. Teils liefern sie zu den jeweiligen Themen didaktisch gut gegliederte Übersichten unter Einbeziehung neuester Ergebnisse, teils sind sie auf die Zusammenfassung der Ergebnisse reduziert. Alle gewähren sie Einblicke in aktuelle Entwicklungen, so wie es der Titel verspricht. In der Einleitung ist von der „explosionsartigen" Entwicklung der Biotechnologie die Rede, aber auch von der Notwendigkeit zur Versachlichung und zur Absage an Hoffnungen auf schnelle, billige Erfolge. Gerade die neuen Gebiete, wie Gentechnik und immobilisierte Biokatalysatoren können nur bei intensiver Bearbeitung fruchtbar werden. Diese Überzeugung vermittelt der Herausgeber in seinem Vorwort ebenso wie der mitgeteilte Stand der Fakten.

K. Sattler

Acta Biotechnol. 5 (1985) 3, 293—296

Wirkung von Ölen und Fettsäuren auf Wachstum und Enzymbildung bei Thermoactinomyces vulgaris

V. Untersuchungen zur Metabolisierung des Öles

LEUCHTENBERGER, A. und RUTTLOFF, H.

Akademie der Wissenschaften der DDR
Forschungszentrum für Molekularbiologie und Medizin
Zentralinstitut für Ernährung
DDR-1505 Potsdam-Rehbrücke, Arthur-Scheunert-Allee 114—116

Summary

The splitting of vegetable and animal oils used as antifoam agents and the metabolization of fatty acids released were analysed during fermentation of *T. vulgaris* in a 30 l jar fermenter. During cultivation time of 22 hours the oil amount decreased by 60 to 70%. The metabolization of fatty acids is 50 to 90% of the starting level depending on the kind of acid. The eventual effect of the fatty acids released from antifoam oils on the protease production is discussed.

Einleitung

Es gibt im Schrifttum verschiedene Hinweise dafür, daß als Entschäumer zugesetzte pflanzliche und tierische Öle und Fette von Mikroorganismen abgebaut und z. T. verstoffwechselt werden [1]. So ermitteln z. B. YAMAMOTO et al. [2], daß diesbezüglich genutzte Öle von *Endomyces* sp. umgesetzt werden und die freigesetzten Fettsäuren für die Anregung der Amylaseproduktion verantwortlich sind. FRANZKE et al. [3] beobachten eine nahezu vollständige Metabolisierung von verschiedenen pflanzlichen Entschäumerölen durch *Endomycopsis bispora.* Auch GALYNKIN et al. [4] finden, daß bei der Kultivierung von *Streptomyces hygroscopicus* Walöl verstoffwechselt wird. Von den freien Fettsäuren werden insbesondere die ungesättigten als C-Quelle genutzt.
Die vorliegende Mitteilung beinhaltet analytische Untersuchungsergebnisse über Veränderungen der als Entschäumer verwendeten pflanzlichen und tierischen Öle durch *T. vulgaris.*

Material und Methoden

Die Versuche werden mit dem Stamm *T. vulgaris* durchgeführt, dessen Haltung kurzzeitig (bis 4 Wochen) auf Maisquellwasser (MQW) I-Agar, langfristig (über Monate) in Form von Sporenkonserven mit Luvos-Heilerde [5, 1] erfolgt. Die Zusammensetzung der verwendeten Medien (MQW I-Agar, MYP-Medium, MQW II-Medium), das Ansetzen von Vor- und Arbeitskultur sowie die Durchführung der Fermentation in 30 l-Fermentoren entsprechen den Angaben in Mitteilung III [6]. Die analytischen Untersuchungen zum Fettabbau sind vom Bereich Lebensmittelchemie der Sektion Chemie an der Humboldt-Universität Berlin durchgeführt worden. Die Fettbestimmung erfolgt nach SOXHLET. Freie Fettsäuren werden mit 0,1 N Kalilauge titrimetrisch erfaßt. Quantitative Veränderungen der Fettsäurezusammensetzung werden gaschromatographisch ermittelt [3].

Ergebnisse

In einer Versuchsserie im 30 l-Glasfermentor werden unter Verwendung von MQW II-Medium jeweils 0,2% Raps-, Soja- bzw. Fischöl zum Zeitpunkt t_0 in das Kulturmedium gegeben [6]. Zu Beginn und zum Abschluß der Fermentation (nach 22 h) werden jeweils 1000 ml Kulturmedium entnommen. Um Fettverluste zu vermeiden, erfolgt ohne vorherige Konzentrierung eine sofortige Lyophilisation. Das getrocknete Gut wird möglichst quantitativ gewonnen und für die nachfolgenden analytischen Untersuchungen verwendet. Der Gehalt an Fett sowie die freien Fettsäuren sind in Tab. 1 zusammengefaßt. Danach sind die in den Kontrollen ohne Ölzusatz vorkommenden Fettmengen zu vernachlässigen. Bei den Ölvarianten nimmt der Fettgehalt während der 22 h dauernden Fermentation um 60 bis 70% ab und gleichzeitig die Menge an freien Fettsäuren erheblich zu.

Tabelle 1. Veränderung der Konzentration an Öl und freien Fettsäuren während der Fermentation im 30 l-Fermentor (MQW II-Medium, 22 h Kultivierung, 0,2% Ölzusatz)

Nr.	Variante	Zeitpunkt d. Proben [h]	Fettgehalt [%]	freie Fettsäuren [%]
1	Kontrolle	0	0,015	—
2		22	0,013	—
3	0,2% Rapsöl	0	0,16	0,1
4		22	0,06	50,3
5	0,2% Sojaöl	0	0,12	0,6
6		22	0,03	32,8
7	0,2% Fischöl	0	0,15	0,7
8		22	0,05	23,0

Tabelle 2: Veränderung der Fettsäurezusammensetzung (in mg) in einem Liter Kulturmedium während der Fermentation im 30 l-Fermentor (MQW II-Medium, 0,2% Ölzusatz, 22 h)

Fettsäure-art	Rapsöl		Abbau	Sojaöl		Abbau	Fischöl		Abbau
	Kultur-dauer [h]			Kultur-dauer [h]			Kultur-dauer [h]		
	0	22	%	0	22	%	0	22	%
14	6,4	29,4		—	23,7		90,0	12,5	86,2
14 : 1	—	—		—	—		4,5	23,5	
16	86,4	9,0	89,5	102,0	17,7	82,6	252,0	47,5	81,2
16 : 1	9,6	12,0		—	4,8		205,5	43,0	79,1
17 : 1	—	—		—	—		7,5	1,5	81,3
18	12,8	1,8	86,0	30,0	4,2	86,0	25,5	8,0	68,7
18 : 1	198,4	60,0	69,7	259,2	64,5	75,1	412,5	125,5	69,6
18 : 2	260,8	73,8	71,7	604,8	115,8	80,8	30,0	34,5	
18 : 3	220,8	60,0	72,8	90,0	14,4	84,0	—	—	
20	—	—		—	—		—	4,0	
20 : 1	—	—		—	—		226,5	83,0	63,4
20 : 3	—	—		—	—		174,0	105,0	29,7
20 : 4	—	—		—	—		72,0	12,0	83,3
22 : 1	804,8	354,0	56,0	114,0	54,9	51,8	—	—	
gesamt	1600,0	600,0		1200,0	300,0		1500,0	500,0	

Die qualitativen und quantitativen Veränderungen in der Fettsäurezusammensetzung während der Fermentation wurden gaschromatographisch verfolgt (Tab. 2). Die Fettsäurebilanzierung bei drei Ölen unterschiedlicher Zusammensetzung zeigt, daß gesättigte wie auch ungesättigte Fettsäuren verstoffwechselt werden. Wenn auch der Abbau der Fettsäuren in unterschiedlichem Grade erfolgt, ist eine ausgesprochene Bevorzugung einer bestimmten Säure nicht erkennbar. Bekanntlich ist der Anteil an Erucasäure im Rapsöl sehr hoch. Im Gegensatz zu allen anderen Fettsäuren, die zu etwa 70 bis 90% verstoffwechselt werden, wird diese nur zu ca. 50% umgesetzt. Die Anwesenheit von Erucasäure im Sojaöl deutet auf Verunreinigungen mit Rapsöl hin.

Diskussion

Die analytischen Untersuchungen im Verlaufe der Fermentation weisen aus, daß das als Entschäumer eingesetzte pflanzliche oder tierische Öl nicht unverändert bleibt. Gemäß Tab. 1 wird dieses zu etwa 60 bis 70% abgebaut, wobei gleichzeitig die Konzentration an freien Fettsäuren im Verlaufe der Kultivierung ganz erheblich zunimmt. Dabei werden Beziehungen zwischen dem Abbaugrad der verschiedenen Öle und der Anreicherung freier Fettsäuren einerseits sowie der Höhe der Proteasesynthese andererseits sichtbar [7]. Mit zunehmendem Ölabbau und geringerer Fettsäureanreicherung im Medium steigt die Proteaseaktivität an. Fischöl ist diesbezüglich am wirkungsvollsten. Es wird zu 66% abgebaut, die Anreicherung freier Fettsäuren beträgt nur 25% und die Zunahme der Proteasesynthese gegenüber der Kontrolle ist mit 60% am höchsten. Bei Rapsöl dagegen beträgt der Abbau weniger als 60%, die Anreicherung freier Fettsäuren liegt bei 50% und die Steigerungsrate der Proteasesynthese bei 40%. Sojaöl liegt in der Mitte. Die Ergebnisse sprechen dafür, daß Art und Menge freiwerdender Fettsäuren während des Fermentationsprozesses für die unterschiedliche Wirkung der verschiedenen Öle verantwortlich sind.

Derartige Befunde sind auch von anderen Autoren bekannt. So ermitteln z. B. YAMAMOTO et al. [2], daß das wirksame Prinzip bei der Anregung der Amylaseproduktion von *Endomyces* sp. durch verschiedene tierische und pflanzliche Entschäumeröle auf Fettsäuren zurückzuführen ist. Glycerin zeigt hingegen keine Wirkung. Von einer Verstoffwechselung des Entschäumeröles in erheblichem Umfang durch *Endomycopsis bispora* berichten auch FRANZKE et al. [3]. Sie finden, daß Rapsöl zu 86 bis 92% und Maiskeimöl zu 97% im Verlaufe von 72 h umgesetzt werden. RQ-Werte von 0,7 bis 0,8 nach der 25. Fermentationsstunde sprechen für eine bevorzugte Fettsäureveratmung. Diese ist bei Maiskeimöl intensiver als bei Rapsöl, was sich in einer um 50% niedrigeren Anreicherungsrate der freigesetzten Fettsäuren bemerkbar macht. Durch gaschromatographische Analysen des Fettsäurespektrums im Kulturmedium weisen diese Autoren jedoch nach, daß die Metabolisierung der verschiedenen Fettsäuren unterschiedlich ist. Während die Abbaurate von Erucasäure mit 80% am geringsten ist, werden andere ungesättigte Fettsäuren durchschnittlich zu 90% und Polyenfettsäuren sogar zu 97% umgesetzt. Gesättigte Fettsäuren weisen im Durchschnitt eine etwas geringere Abbaurate auf.

Die gaschromatographische Analyse des Fettsäurespektrums im Medium bei Kultivierung von *T. vulgaris* führt zu andersartigeren Ergebnissen. Generell liegt die Abbaurate der meisten Fettsäuren im Bereich zwischen 70 und 90%. Eine Ausnahme — ähnlich wie von FRANZKE et al. [3] beschrieben — bildet Erucasäure, die nur zu 50 bis 60% verstoffwechselt wird. Die bessere Wirkung von Sojaöl und Fischöl — im Vergleich zu Rapsöl — auf die Proteasesynthese könnte eventuell mit dem geringeren bzw. fehlenden Gehalt dieser Säure in Beziehung stehen. Eine bevorzugte Metabolisierung ungesättigter Fettsäuren, wie sie von YAMAMOTO et al. [2], GALYNKIN et al. [4] sowie FRANZKE et al. [3] nachgewiesen wird, ist bei *T. vulgaris* nicht erkennbar.

Daß der Ausnutzungsgrad der Fettsäuren allein noch keine Rückschlüsse auf ihre aktivitätsfördernde Wirkung zuläßt, geht aus Untersuchungen von YAMAMOTO et al. [2] hervor: Obwohl ungesättigte Fettsäuren schneller als gesättigte abgebaut werden, wird bei ersteren offenbar kein wesentlicher Einfluß auf die Steigerung der Enzymaktivität gefunden. Eine solche wird vor allem von einem Fettsäuregemisch angeregt, in dem gesättigte Fettsäuren dominieren. Palmitinsäure spielt dabei eine besondere Rolle. Zu ähnlichen Erkenntnissen gelangen auch YAKOVLEVA et al. [8] bei Versuchen mit *T. vulgaris* in einem synthetischen Medium. Die Zugabe von einzelnen, in den 3 getesteten Ölen dominierenden Fettsäuren zu dem Grundmedium ergibt nur bei Palmitinsäure einen Stimulationseffekt innerhalb von 24 h. Dagegen zeigen alle ungesättigten Säuren eine mehr oder weniger große Hemmwirkung. Diese ist, in Übereinstimmung mit Ergebnissen von ANDERS et al. [9], GALBRAITH et al. [10] sowie FREESE et al. [11], vor allem auf eine Wachstumshemmung zurückzuführen. Dabei spielt die Konzentration der frei vorliegenden Fettsäuren eine wichtige Rolle. Wie die Arbeit ausweist, geht bei sehr niedrigen Konzentrationen auch der Hemmeffekt ungesättigter Fettsäuren zurück. Der Schwellenwert ist für die einzelnen Fettsäuren sehr verschieden. Während er für Ölsäure, Linol- und Linolensäure besonders niedrig liegt (0,01 bis 0,02%), befindet er sich bei Erucasäure zwischen 0,05 und 0,1%. Glycerin regt zwar das Wachstum an, bewirkt aber — ähnlich wie bei YAMAMOTO et al. [2] — keine Aktivitätserhöhung. Bei kombinierter Anwendung einzelner Fettsäuren sowie von Fettsäuren mit Glycerin erhalten YAKOVLEVA et al. [12] unterschiedliche Befunde. Während der Einsatz gemischter Fettsäuren generell zu einer Hemmung der Enzymsynthese führt, wirken sowohl Palmitin- als auch Erucasäure in Verbindung mit Glycerin aktivitätsfördernd. Das Ergebnis ist bei Erucasäure überraschend, da diese Säure allein in gleicher Konzentration keine Stimulationswirkung zeigt. Die durch Zusatz von Palmitin- bzw. Erucasäure zu Rapsöl erzielte weitere Aktivitätserhöhung ist sicherlich durch die sofortige Nutzung dieser Säuren in der Wachstumsphase zu erklären, bevor durch Lipolyse auch die aus dem Öl freigesetzten Fettsäuren verfügbar sind. Zu ähnlichen Befunden kommen auch HIRAGA et al. [13], indem sie 0,5% Fettsäureester (vorzugsweise Stearat und Palmitat) zu 0,5% pflanzlicher Öle zusetzen und damit eine weitere Steigerung der Peptidasesynthese durch *Aspergillus sojae* und *Aspergillus oryzae* erreichen.

Eingegangen: 26. 3. 1984

Literatur

[1] LEUCHTENBERGER, A., KLINGENBERG, P., RUTTLOFF, H.: Z. Allg. Mikrobiol. **19** (1979), 27.
[2] YAMAMOTO, T., HATTORI, F., TAKATSU, H.: Bull. Inst. Chem. Res. Kyoto **42** (1964), 252.
[3] FRANZKE, C., GÖBEL, R., KROLL, J., LIEBS, P., RITTER, G., SCHULTZE, M.: Nahrung **23** (1979), 283.
[4] GALYNKIN, V. A., EFIMOVA, T. P., GREKOVA, V. V., KOROLEVA, T. A.: Khim.-Farm. Zh. **12** (1978), 116.
[5] KLINGENBERG, P., ZICKLER, F., LEUCHTENBERGER, A., RUTTLOFF, H.: Z. Allg. Mikrobiol. **19** (1979), 17.
[6] LEUCHTENBERGER, A., RUTTLOFF, H.: Z. Allg. Mikrobiol. **23** (1983), 644.
[7] LEUCHTENBERGER, A., RUTTLOFF, H.: Z. Allg. Mikrobiol. (im Druck)
[8] YAKOVLEVA, M., LEUCHTENBERGER, A., RUTTLOFF, H., LOGINOVA, L. G.: Mikrobiologija (im Druck)
[9] ANDERS, R. F., JAGO, G. R.: J. Dairy Res. **31** (1964), 81.
[10] GALBRAITH, H., MILLER, T. B., PATON, A. M., THOMSON, J. K.: J. Appl. Bacteriol. **34** (1971), 803.
[11] FREESE, E., SHEU, C. W., GALLIERS, E.: Nature **241** (1973), 321.
[12] YAKOVLEVA, M., LEUCHTENBERGER, A., RUTTLOFF, H., LOGINOVA, L. G.: Mikrobiologija (im Druck)
[13] HIRAGA, H., MIYAJIMA, R., MITSUGI, K.: DB-OS 2813247.

Acta Biotechnol. 5 (1985) 3, 297—308

Analytik atypischer Fettsäuren in biologischen Matrizes

PÖRSCHMANN, J.[1], PÖRSCHMANN, S.[2], LIEBETRAU, L.[1] und RICHTER, K.[2]

[1] Institut für technische Mikrobiologie
DDR-1017 Berlin, Alt-Stralau 62
[2] Akademie der Wissenschaften der DDR
Institut für Biotechnologie Leipzig
DDR-7050 Leipzig, Permoserstraße 15

Summary

By combined application of chemical pretreatments, capillary gas-chromatography and mass spectrometry it was possible to enlighten the structure of atypical fatty acids with hydroxy groups and cyclopropane rings under the use of only a few of reference substances.
The direct alkaline saponification of the sample with liberation of fatty acids and following methylation with boron trifluoride/methanol or diazomethane was proved to be the best method regarding to precision and speed of the sample cleanup.

Einleitung

Für die Führung biochemischer Prozesse sowie für die Qualitätskontrolle mikrobiell erzeugter Produkte gewinnt die Untersuchung der Fettsäurezusammensetzung der untersuchten Matrizes steigendes Interesse [1].
Als Fettsäuren werden gewöhnlich gesättigte und ungesättigte Karbonsäuren mit einer Kohlenstoffkette von ca. 12—22 C-Atomen bezeichnet [2]. Heutzutage ist jedoch bekannt, daß z. B. über den Stoffwechsel von Bakterienkulturen Karboxylverbindungen mit Kettenverzweigungen, Doppelbindungen in trans-Stellung, Hydroxy-, Oxo-Cyclopropan- und Cyclopropenstrukturen entstehen können. Diese Spezies bezeichnet man als atypische oder seltene Fettsäuren.
Zur Lipiduntersuchung komplex zusammengesetzter Proben verwendet man heutzutage vor allem chromatographische Verfahren. Während sich die Dünnschicht-Chromatographie (DC) zur Vortrennung von Lipidklassen durchgesetzt hat, wird zur Trennung und Identifizierung der Fettsäureindividuen vor allem die Gas-Chromatographie (GC) mit hochtemperaturstabilen flüssigen stationären Phasen angewandt.
In diesem Beitrag soll gezeigt werden, wie durch Kombination von prä-chromatographischen, kapillar-gas-chromatographischen (KGC) und spektroskopischen Trenn- und Identifizierungstechniken ein komplex zusammengesetztes Fettsäuregemisch nahezu vollständig qualitativ aufgeklärt wurde.

Experimentelles

Aufarbeitung der Probe und prä-chromatographische Vorbehandlungen

Die Aufarbeitung der Probe bis zu den Fettsäuremethylestern (FSME) sowie die chemische Vorbehandlung des FSME-Gemisches sind in Abb. 1 dargestellt. Eine ausführliche Beschreibung der Arbeitsgänge Extraktion, Verseifung, Freisetzung der Fettsäuren und Veresterung sowie der prä-chromatographischen Vorbehandlungen wie Silylierung, Hydrierung und Bromierung ist in [3] angegeben. Die Fixierung von Oxo-Fettsäuren wird in [4] detailliert beschrieben.

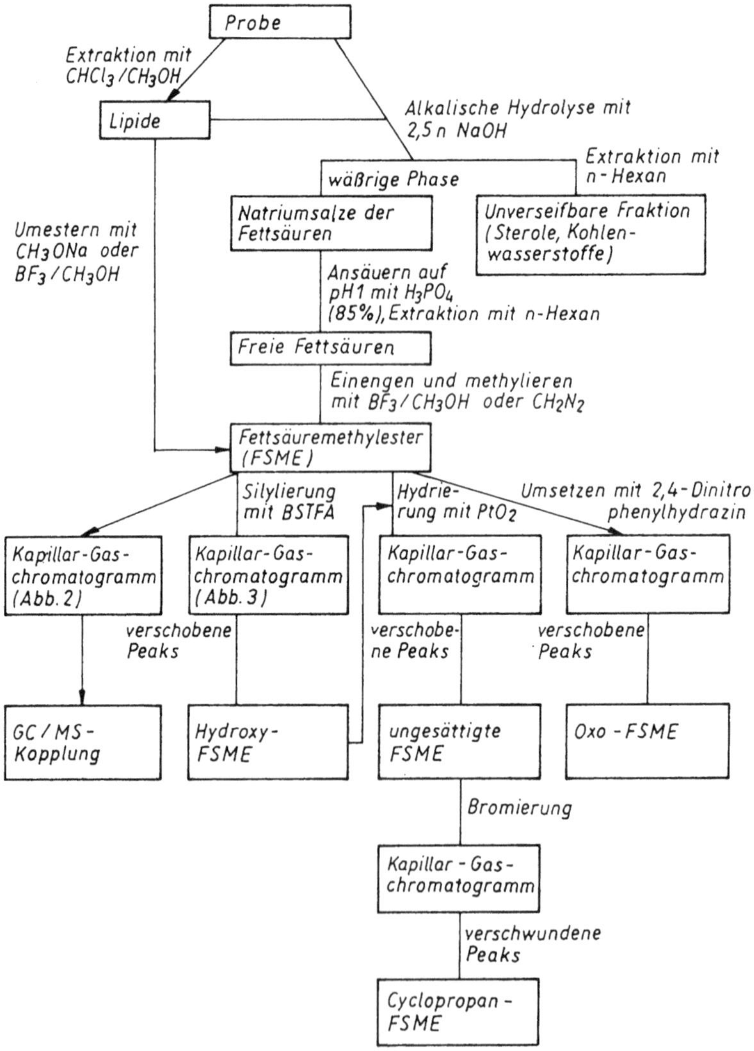

Abb. 1. Probenaufbereitung und prä-chromatographische Vorbehandlung

Die Darstellung von Methylestern mittels Diazomethan erfolgte in einem von FALES [5] beschriebenen Reaktionsgefäß, wobei das Veresterungsreagens aus N-Nitroso-N-methyl-p-toluen-sulphonamid freigesetzt wurde.

Kapillar-Gas-Chromatographie und Kopplung mit der Massenspektroskopie (GC/MS- Kopplung)

Geräte: Kapillar-Gas-Chromatograph HP 5840 A (Hewlett Packard/USA), GC/MS-Gerät Finnigan 4021 (Finnigan/USA)

Kapillaren:
- 21 m Länge · 0,22 mm innerer Durchmesser, statisch imprägniert mit OV 1
- 60 m Länge · 0,25 mm innerer Durchmesser, dynamisch imprägniert mit Carbowax 20 M

Analysenbedingungen siehe Abb. 2 und 3.

Resultate und Diskussion

Probenaufbereitung

Das Ziel der Lipidextraktion besteht darin, die zellulären Lipide in einem unzerstörten Zustand und nichtkontaminiert mit nichtlipidischen Bestandteilen wie z. B. Kohlenhydraten oder Proteinen zu gewinnen. Dabei hängt der Erfolg einer Extraktion ab von der chemischen Natur der Lipide und der Art ihrer Verbindung in den Zellen. Entsprechend der ausgeübten Wechselwirkungen von Lipiden lassen sie sich prinzipiell in drei Typen einteilen [6] (Tab. 1).

Tabelle 1. Einteilung von Lipiden

Lipidklasse	Vertreter	Extraktion mit
„neutrale“ oder unpolare Lipide	Glyzeride, Sterolester, Kohlenwasserstoffe	unpolaren Lösungsmitteln wie Chloroform, Benzen, Diethylether
polare Lipide	Phosphatide, Glycolipide	polaren Lösungsmitteln wie Methanol oder Ethanol
kovalent gebundene Lipide	Lipopolysaccharide	—

Die kovalent gebundenen Lipide, in denen z. B. Hydroxy-Fettsäuren ester- oder amidartig an andere zelluläre Komponenten gebunden sind, müssen vom Komplex durch eine Verseifung entfernt werden. Auf Grund der Vollständigkeit der Lipidextraktion als auch auf Grund der wenig zeitaufwendigen Durchführbarkeit bevorzugten wir gegenüber einer FOLCH-Extraktion mit Chloroform/Methanol (2 : 1) eine direkte alkalische Verseifung der Probe.

Wie aus Abb. 1 zu erkennen ist, besteht eine weitere, sehr einfache und zeitsparende Methode der Herstellung von FSME in der Umesterung, die gewöhnlich mit BF_3 [7] oder CH_3ONa katalysiert wird. (Bei Anwendung von basischen Alkalialkoholaten werden die freien Fettsäuren nicht erfaßt).

Ein Nachteil der Methode besteht darin, daß die unverseifbare Fraktion nicht entfernt wird und gegebenenfalls im Chromatogramm stört. Außerdem werden die kovalent gebundenen Fettsäuren nicht erfaßt [8].

Für eine gute Reproduzierbarkeit und Vergleichbarkeit der Analysenergebnisse ist nicht nur die Art der Isolierung der Fettprobe, sondern auch die Veresterung der Fettsäuren von Bedeutung. Die Notwendigkeit einer Veresterung — für Fettsäuren mit C-Zahlen zwischen 12 und 24 wird eine Methylierung bevorzugt — ergibt sich einerseits aus der rasch abnehmenden Flüchtigkeit innerhalb der homologen Reihe der Fettsäuren (Siedepunkt Ameisensäure: 100 °C, Siedepunkt Kapronsäure: 205 °C) sowie andererseits aus deren hoher Polarität. Beide Faktoren wirken einer schnellen, diskriminierungsfreien Elution von der gas-chromatographischen Säule entgegen.
Von den verschiedenen Methylierungsvarianten haben sich die BF_3- und die Diazomethan-Methode auf Grund der hohen Veresterungsrate von 95—98% und der kurzen Reaktionszeit durchgesetzt. Wir wandten beide Methoden an und erhielten im C-Zahl-Bereich größer 10 absolut identische Kapillar-Gas-Chromatogramme.

Identifizierung der Fettsäureindividuen mittels prä-chromatographischer Modifizierungen

Wie aus Abb. 2 zu erkennen ist, läßt sich das komplex zusammengesetzte Fettsäuremethylestergemisch auf einer hocheffizienten und gut desaktivierten OV 1-Kapillare nahezu vollständig in die Individuen auflösen. Mit Hilfe zur Verfügung stehender Testsubstanzen konnten wir in der von uns untersuchten Probe von Bakterienlipiden Myristin-, Palmitin-, Stearin-, Öl- und Linolsäure nachweisen. Weitere wichtige Identifizierungshilfen stellen die prä-chromatographischen Vorbehandlungen dar.
Durch diese Modifizierungen der nativen Probe ergab sich eine Möglichkeit zur Klassifizierung nach funktionellen Gruppen von kapillar-gas-chromatographisch aufgetrennten Fettsäuremethylesterindividuen. In unserem Fall wird ein spezifisches, auf die betreffende funktionelle Gruppe (z. B. Hydroxygruppe oder Cyclopropanring) selektiv wirkendes Reagens verwendet (Silylierungsmittel bzw. Brom (Abb. 1). Nach chemischer Vorbehandlung kann man dann gegenüber dem Chromatogramm der nativen Probe im Chromatogramm der chemisch modifizierten Probe Veränderungen wie Peakverschiebungen, Entfernung oder Neubildung von Peaks beobachten und auf dieser Basis Zuordnungen zu Fettsäuretypen treffen.
Die angewandte Methodik der Zuordnung von Fettsäuretypen soll am Beispiel der Hydroxy-Fettsäuren näher erläutert werden. Durch Vergleich der unter identischen Analysenbedingungen aufgenommenen Chromatogramme von nativer und silylierter Probe (Abb. 2 und 3) konnte geschlußfolgert werden, daß 6 Hydroxy-Fettsäuren im Gemisch enthalten sind. Die Peaks mit der Gesamtretentionszeit von 21,19; 33,21; 34,09; 38,97; 43,67 und 46,35 min (in Abb. 2 zur besseren Übersicht mit der Numerierung 1—6 versehen) sind nach der Silylierung (Abb. 3) entweder verschwunden oder in ihrer Größe deutlich vermindert. In Abb. 3 treten gegenüber Abb. 2 bei höheren Retentionszeiten neue Peaks auf: Die Silylierungsprodukte (in Abb. 3 mit 1′—6′ versehen).
Durch eine Hydrierung der nativen bzw. silylierten Probe ergab sich weiterhin, daß der Peak bei 43,67 (Peak 5 in Abb. 2) bzw. dessen Silylierungsprodukt (Peak 5′ in Abb. 3) seine Position im Chromatogramm verändert, d. h. es handelt sich um eine ungesättigte Hydroxy-Fettsäure. Wird danach die silylierte/hydrierte Probe einer Bromierung unterworfen, verschwindet der Peak 6′ (Abb. 3). Daraus läßt sich der Schluß ziehen (Abb. 1), daß es sich bei der Fettsäure 6 (Abb. 2) um eine Hydroxy-Cyclopropan-Verbindung handelt. Die übrigen 4 Peaks der Silylierungsprodukte verändern weder nach Hydrierung noch Bromierung ihre Lage im Chromatogramm, d. h. es sind 4 gesättigte Hydroxy-Fettsäuren im Lipidextrakt enthalten.
In analoger Weise konnten durch Hydrierung weitere 3 ungesättigte Verbindungen und durch Kombination von Hydrierung und Bromierung bzw. Kochen der nativen Probe mit methanolischer Salzsäure 2 weitere Cyclopropan-Fettsäuren nachgewiesen werden. Die Unterscheidung von monoungesättigten und Cyclopropan-Fettsäuren mittels

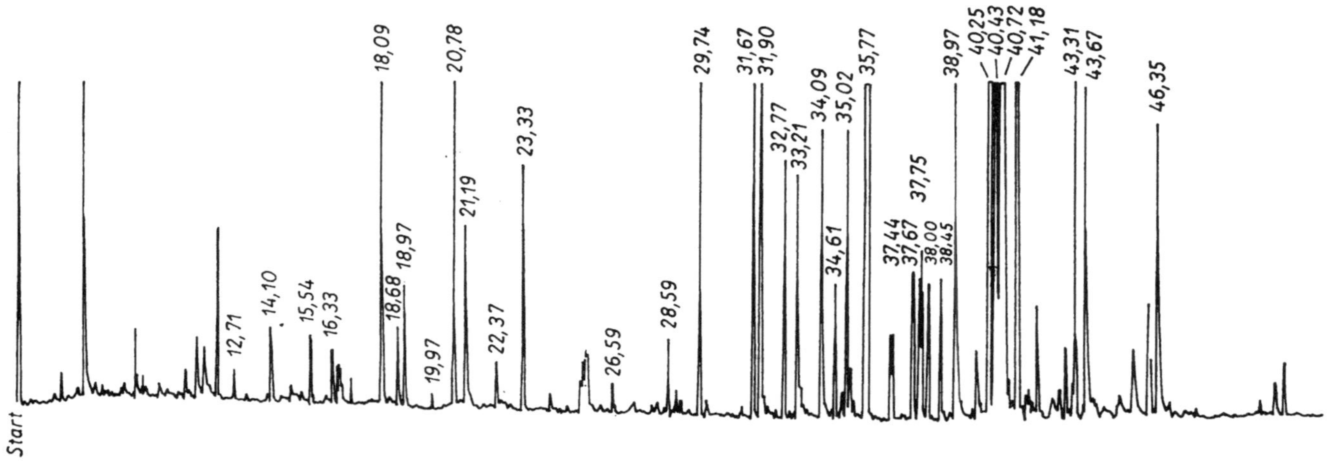

Abb. 2. Kapillar-Gas-Chromatogramm der nativen Probe auf OV 1

TEMP 1	400	40	40	FID SGNL	+ B	
TIME 1	0,00			SLP SENS	0,60	
RATE	3,50			AREA RE I	1000000000	
TEMP 2	400	220		FLOW A	0,0	0,2
TIME 2	2,00			FLOW B	0,0	0,2
FIN TEMI	400	230	230			
AUX TEMP	400	260	260	0,01 VLV/EXT	1	
				0,50 VLV/EXT	1	
CHT SPD	0,10			4,00 AREA RE I	2000	
ZERO	5,0			5,00 CHT SPD	1,20	
ATTN 2↑	4					

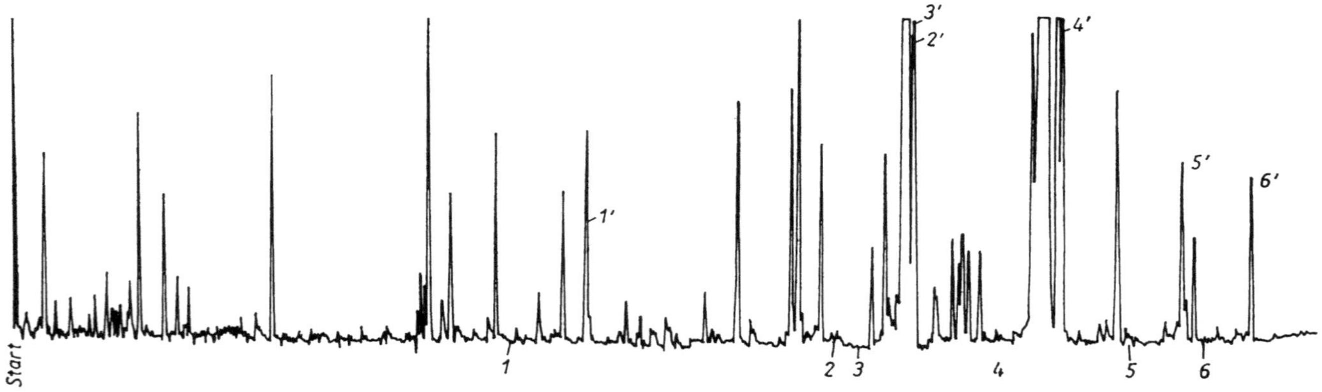

Abb. 3. Kapillar-Gas-Chromatogramm der silylierten Probe auf OV 1

TEMP 1	400	40	40	SLP SENS	0,60	
TIME 1	0,00			AREA RE I	1000000000	
RATE	3,50			FLOW A	0,0	0,3
TEMP 2	400	240		FLOW B	0,0	0,2
TIME 2	10,00					
FID TEMP	400	240	240	0,02 VLV/EXT	1	
AUX TEMP	400	260	260	0,05 IFN	10	
				0,40 VLV/EXT—	1	
CHT SPD	0,02			9,50 AREA RE I	2000	
ZERO	8,0			10,00 CHT SPD	1,20	
ATTN 2↑	2			58,00 STOP		
FID SGNL	+ B					

prä-chromatographischer Methoden ist von besonderem Interesse, da sie sich auf Grund ihres ähnlichen Retentionsverhaltens und ihrer ähnlichen massenspektroskopischen Fragmentierung nur schwer gas-chromatographisch bzw. massenspektroskopisch unterscheiden lassen (siehe die folgenden beiden Abschnitte). Auf die Abwesenheit von Oxo-Fettsäuren konnte geschlossen werden, da die Chromatogramme von nativer und mit 2,4-Dinitrophenylhydrazin umgesetzter Probe absolut identisch waren. Eine Zusammenfassung aller durch prä-chromatographische Verfahren gewonnenen Informationen vermittelt Tab. 2.

Tabelle 2. Identifizierung von Fettsäuremethylestern

Gesamt-retentionszeit (Abb. 2)	ECL (OV 1) 200 °C	ECL (Carbowax 20 M) 190 °C	ΔECL	Informationen aus				Gesamtaussage
				Hydrierung	Bromierung	Silylierung	Struktur-Retentions-Beziehungen	
12,71	9,00	9,00	0,00	g			9C, g, n	Nonansäure
14,10	9,42							
15,54	9,80	9,78	−0,02	g			10C, g, verzw.	
16,33	10,00	10,00	0,00	g			10C, g, n	Dekansäure
18,09	10,51							
18,68	10,65	10,60	−0,05	g			11C, g, iso	iso-Undekansäure
18,97	10,73	10,72	−0,01	g			11C, g, anteiso	anteiso-Undekansäure
19,97	11,00	11,00	0,00	g			11C, g, n	Undekansäure
20,78	11,26							
21,19	11,39	—	—	g		h	10C	3-Hydroxy-Dekansäure
22,37	11,73	11,71	−0,02	g			12C, g, anteiso	anteiso-Dodekansäure
23,33	12,00	12,00	0,00	g			12C, g, n	Dodekansäure
26,59	13,00	13,00	0,00	g			13C, g, n	Tridekansäure
28,59	13,46							
29,74	14,00	14,00	0,00	g			Testsubstanz	Tetradekansäure
31,67	14,64	14,58	−0,06	g			15C, g, iso	iso:Pentadekansäure
31,90	14,72	14,72	0,00	g			15C, g, anteiso	anteiso-Pentadekansäure
32,77	15,00	15,00	0,00	g			15C, g, n	Pentadekansäure
33,21	15,12	—	—	g		h	14C	2-Hydroxy-Tetradekansäure
34,09	15,40	—	—	g		h	14C	3-Hydroxy-Tetradekansäure

Tabelle 2 (Fortsetzung)

Gesamt-retentionszeit (Abb. 2)	ECL (OV 1) 200 °C	ECL (Carbowax 20 M) 190 °C	ΔECL	Information aus				Gesamtaussage
				Hydrierung	Bromierung	Silylierung	Struktur-Retentions-Beziehungen	
34,61	15,64	15,59	−0,05	g			16 C, g, iso	iso-Hexadekansäure
35,02	15,76	16,15	0,39	ug			$C_{16:1\omega7c}$	Palmitoleinsäure
35,77	16,00	16,00	0,00	g			Testsubstanz	Palmitinsäure
37,44	16,64	16,60	−0,04	g			17 C, g, iso	iso-Heptadekansäure
37,67	16,73	16,71	−0,02	g			17 C, g, anteiso	anteiso-Heptandekansäure
37,75	16,76	17,12	0,36	ug			17 C, monoungesättigt	monoungesättigte Heptadekansäure
38,00	16,85	17,25	0,40		c		17 C	9,10-Methylen-Palmitinsäure
38,45	17,00	17,00	0,00	g			17 C, g, n	Heptadekansäure
38,97	17,20	—	—	g		h	16 C, g	2-Hydroxy-Hexadekansäure
40,25	17,63	18,47	0,86	ug			Testsubstanz	Linolsäure
40,43	17,69	18,11	0,42	ug			Testsubstanz	Ölsäure
40,72	17,76	18,13	0,37	ug			$C_{18:1\omega7c}$	cis-Vaccensäure
41,18	18,00	18,00	0,00	g			Testsubstanz	Stearinsäure
43,31	18,75	19,24	0,39		c		19 C	11,12-Methylen-Oktadekansäure
43,67	18,81	—	—	ug		h	18 C	2-Hydroxy-cis-Vaccensäure
46,35	19,82	—	—		c	h	19 C	2-Hydroxy-11,12-Methylen-Oktadekansäure

Symbole: g — gesättigt
ug — ungesättigt
h — Hydroxy-
c — Cyclopropan-

Gas-chromatographische Identifizierung

Die Verwendung von reproduzierbaren Retentionsdaten ist eine sehr schnelle und zuverlässige Identifizierungsmethode für langkettige FSME.
An unpolaren stationären Phasen wie OV 1 werden die gas-chromatographischen Retentionsdaten gemäß der fundamentalen HERINGTON-Trennformel [10] vor allem durch die Siedepunkte der chromatographierten Verbindungen bestimmt. Bei isothermer Arbeitsweise folgt für homologe Reihen die bekannte lineare Abhängigkeit des logarithmierten Retentionsvolumens (lg V_g) von der C-Zahl der chromatographierten Verbindung. Das Ziel unserer Arbeiten bestand darin, auf der Basis dieser und analoger Struktur-Retentions-Beziehungen (s. u. und [3]) getrennte Fettsäuremethylesterindividuen bezüglich ihrer C-Zahl, der Art der Kette (verzweigt, unverzweigt) und bei ungesättigten Komponenten nach dem Grad der Sättigung unter Nutzung der durch die prä-chromatographischen Modifizierungen gewonnenen Informationen weiter zu charakterisieren.
Zur Charakterisierung der Retention verwendeten wir die äquivalenten Kettenlängen*. Unter Verwendung der Testsubstanzen Myristin-, Palmitin- und Stearinsäure konnten wir ein lg V_g-C-Diagramm aufnehmen und auf dieser Grundlage durch Interpolation die Homologen Dekan-, Undekan-, Dodekan (Laurin)-, Tridekan-, Pentadekan- und Heptadekansäuremethylester zuordnen. Damit war die Grundlage für eine exakte ECL-Bestimmung der übrigen Komponenten gegeben (Tab. 2). Die analoge Prozedur wurde zwecks Absicherung der Zuordnung der homomorphen FSME auf der zur Verfügung stehenden polaren Carbowax 20 M-Kapillare durchgeführt (ECL-Werte auf Carbowax 20 M (Tab. 2)**.
Durch Differenzbildung der ECL-Werte auf Carbowax 20 M und OV 1 (ΔECL in Tab. 2) konnten weiterhin Aussagen über den Grad der Sättigung der chromatographierten Individuen getroffen werden:

- Die beiden ungesättigten Komponenten (Gesamtretentionszeit 35,02 und 40,72 (Abb. 2.) besitzen einen ΔECL-Wert von 0,39 und 0,37 und sind gemäß Literaturangaben [12, 13] monoungesättigte Verbindungen.
- Die als Testsubstanz zur Verfügung stehende Linolsäure ist die einzige im Lipidextrakt vorkommende diungesättigte Fettsäure. Polyungesättigte Spezies treten erwartungsgemäß [14] nicht auf.

Trägt man die in der Literatur [13, 15] bekannten ECL_{OV1}-Werte für Verbindungen des Palmitoleintyps (n: 1ω7c) gegen die C-Zahl auf, so lassen sich die von uns gemessenen Werte gut in die Gerade einfügen (eingezeichnete Punkte in Abb. 4), d. h. die beiden ungesättigten Verbindungen (bei 35,02 und 40,72 in Abb. 2) sind Palmitolein- und cis-Vaccensäure.
In analoger Weise kann man unter Verwendung von Literaturdaten [12, 13, 15] für iso- und anteiso- verzweigte gesättigte FSME separate Geraden erhalten, in die sich wiederum von uns gemessene Daten sinnvoll einordnen lassen (Abb. 4). Mit Hilfe dieser grundlegenden Struktur-Retentions-Beziehung konnten wir im Lipidextrakt folgende verzweigte Fettsäuren nachweisen: iso-C_{11}, anteiso-C_{11}, anteiso-C_{12}, iso-C_{15}, anteiso-C_{15}, iso-C_{17}, iso-C_{17} und anteiso-C_{17}.

* Die äquivalenten Kettenlängen (auch ECL-Werte genannt) sind bekanntlich logarithmische Größen, wobei in Analogie zum Retentionsindex-Konzept auf das Netz der homomorphen FSME bezogen wird. Definitionsgemäß besitzt Palmitinsäuremethylester einen ECL-Wert von 16,00, Stearinsäuremethylester von 18,00 usw. Voraussetzung für eine sinnvolle Anwendbarkeit des Konzeptes besteht darin, daß die Nettoretentionszeiten der homomorphen FSME, die als Standard dienen, bekannt sein müssen.

** Auf der Carbowax 20 M-Kapillare war eine ECL-Bestimmung der Hydroxy-Verbindungen nicht möglich, da diese auf der nicht desaktivierten Glasoberfläche adsorbiert werden.

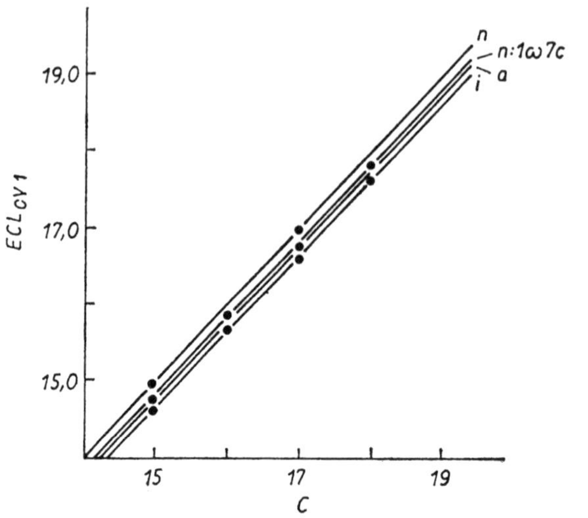

Abb. 4. Struktur- und Retentions-Beziehungen für Fettsäuremethylester auf OV 1

GC/MS-Kopplung

Unter Verwendung von prä-chromatographischen Modifizierungen und gas-chromatographischen Struktur-Retentions-Beziehungen war es möglich, wertvolle Informationen über den Fettsäuretyp, die Kettenlänge, die Art der Kettenverzweigung und den Grad der Sättigung zu gewinnen. Eine vollständige Charakterisierung der atypischen Hydroxy- und Cyclopropanverbindungen mit Angabe der Stellung der funktionellen Gruppe in der Kohlenstoffkette konnte jedoch damit nicht erbracht werden.
Diesbezügliche Angaben sind unter Anwendung der heutzutage wohl leistungsfähigsten analytischen Variante, der Kopplung der Gas-Chromatographie mit der Massenspektroskopie, möglich, da sich beide Analysenmethoden in nahezu idealer Weise ergänzen [16].
Das massenspektroskopische Fragmentierungsverhalten der FSME ist durch die grundlegenden Arbeiten von Ryhage und Stenhagen gut bekannt ([17, 18] und dort zitierte Literatur). Die Interpretation der Massenspektren gestaltet sich jedoch oftmals wegen der vielfältigen Isomeriemöglichkeiten der untersuchten Verbindungen zu einem diffizilen Problem, das auch durch Anwendung eines computergestützten Spektren-Bibliotheksvergleiches nicht immer gelöst werden kann.
Besteht einerseits das Ziel unserer massenspektroskopischen Arbeiten darin, die bereits gewonnenen Informationen zu präzisieren und zu vervollkommnen, so ist andererseits eine exakte und zeitsparende Interpretation der Massenspektren erst unter Verwendung dieser Vorinformationen möglich. Dieser Gedanke kommt (wie bereits oben erwähnt) besonders bei der Interpretation von Massenspektren der Cyclopropan-Verbindungen zum Ausdruck, die ähnlich monoungesättigten FSME fragmentieren [17]. Auf eine explizite Darstellung des massenspektroskopischen Verhaltens der von uns identifizierten atypischen Hydroxy- und Cyclopropan-Verbindungen soll in diesem Rahmen verzichtet werden (Identifizierungshilfen sind in [3] gegeben). Als ein Beispiel sei das Massenspektrum des von uns identifizierten Hydroxy-Cyclopropan-Fettsäuremethylesters diskutiert (Peak 6 in Abb. 2). In Abb. 5 sind charakteristische Fragmente ausgewiesen, die eine Identifizierung dieser Komponente als 2-Hydroxy-11,12-methylen-stearinsäuremethylester ermöglichten.

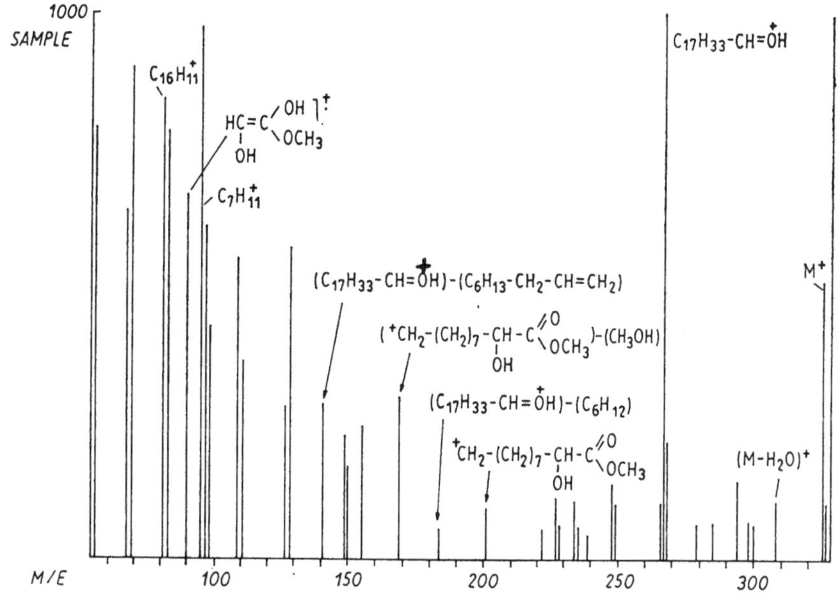

Abb. 5. Massenspektrum von 2-Hydroxy-11,12-methylen-stearinsäuremethylester

In analoger Weise wurden die Hydroxygruppen und Cyclopropanringe bei den betreffenden Verbindungen lokalisiert.
Mit Hilfe der GC/MS-Kopplung konnten weiterhin die mittels gas-chromatographischer Struktur-Retentions-Beziehungen vermuteten Kettenverzweigungen abgesichert werden.
Eine zusammenfassende Gesamtaussage der prä-chromatographisch, gas-chromatographisch und massenspektroskopischen gewonnenen Informationen vermittelt Tab. 2.

Schlußfolgerungen

Unter Anwendung von klassisch chemischen Reaktionen und bekannten Struktur-Retentions-Beziehungen lassen sich in einem komplex zusammengesetzten Fettsäuregemisch die Individuen hinsichtlich Fettsäuretyp, Kettenlänge, Grad der Sättigung und Art der Verzweigung der Kohlenstoffkette auch ohne die anspruchsvolle GC/MS-Kopplung klassifizieren.
Dies ist besonders für Laboratorien von Interesse, die über hocheffiziente Trennsysteme verfügen, jedoch nicht über diese Analysenmethodenkombination.
Für die Fixierung der Stellung der funktionellen Gruppen bei atypischen Fettsäuren scheint die Anwendung der GC/MC-Kopplung unumgänglich, falls nicht in größerem Umfang diese selten vorkommenden Verbindungen als Testsubstanzen zur Verfügung stehen.

Eingegangen: 19. 3. 1984

Literatur

[1] VISWANATHAN, C. V.: Chrom. Rev. **11** (1969), 153.
[2] POHL, P., WAGNER, H.: Fette, Seifen, Anstrichm. **74** (1972), 424, 541.
[3] PÖRSCHMANN, J., WELSCH, T., HERZSCHUH, R., ENGEWALD, W., MÜLLER, K.: J. Chromatogr. **241** (1982), 73.
[4] WEIHRAUCH, J. L., BREWINGTON, C. R., SCHWARTZ, D. P.: Lipids **9** (1974), 883.
[5] DROZD, J.: J. Chromatogr. **113** (1975), 303.
[6] KATES, M.: Techniques of Lipidology. In: Laboratory Techniques in Biochemistry and Molecular Biology, Vol. 3. North Holland Publishing Company, 1973.
[7] HEISZ, O.: G-I-T. Fachz. Lab. **21** (1977), 22.
[8] MOSS, C. W., LAMBERT, M. A., MERWIN, W. H.: Appl. Microbiol. **28** (1974), 80.
[9] REICHMANN, G., JAKOBZA, D., SCHULZE, P.: Z. med. Labortechn. **12** (1971), 233.
[10] LEIBNITZ, E., STRUPPE, H.-G.: Handbuch der Gas-Chromatographie, 2. Auflage. Leipzig: Akad. Verlagsges. Geest u. Portig K.-G., 1984, 46.
[11] ASHES, J. R., HAKEN, J. K., MILIS, S. C.: J. Chromatogr. **187** (1980), 297.
[12] HOFSTETTER, H. H., SEN, N., HOLMAN, R. T.: J. Am. Oil Chem. Soc. **42** (1965), 537.
[13] GOLOVNYA, R. V., URALETS, V. P., KUZMENKO, T. E.: J. Chromatogr. **121** (1976), 118.
[14] THIELE, O. W., BUSSE, D., HOFFMANN, K.: Eur. J. Biochem. **5** (1968), 513.
[15] GOLOVNYA, R. V., KUZMENKO, T. E.: Chromatographia **10** (1977), 545.
[16] REMANE, H., HERZSCHUH, R.: Massenspektrometrie in der organischen Chemie. Berlin: Akademie-Verlag, 1977. 152.
[17] ZEMAN, A., SCHARMANN, H.: Massenspektrometrie von Lipiden; Fette, Seifen, Anstrichm. **74** (1972), 509.
[18] BINNEMANN, P., JAHR, D.: Fresenius Z. Anal. Chem. **297** (1979), 341.
[19] HERZSCHUH, R., PÖRSCHMANN, J.: in Vorbereitung.

Acta Biotechnol. 5 (1985) 3, 309–312

Short Communications

Zur Sorption von Cellulase an chemisch modifizierte Cellulosepulver

PHILIPP, B.[1], KASULKE, U.[1], SCHULZ, G.[2] und HIRTE, W.[2]

[1] Akademie der Wissenschaften der DDR
Institut für Polymerenchemie
DDR-1530 Teltow-Seehof, Kantstraße 55
[2] Humboldt-Universität Berlin
DDR-1086 Berlin, Unter den Linden 10

In früheren Arbeiten [1, 2] hatten wir über die homogene enzymatische Hydrolyse niedrig substituierter, wasserlöslicher Cellulosederivate berichtet und dabei u. a. festgestellt, daß eine Substitution von Hydroxylgruppen am Anhydroglucosering durch anionische Gruppen den enzymatischen Abbau stärker hemmt als eine Derivatisierung zu nichtionischen Ethern (Methylcellulose, Cyanoethylcellulose) mit etwa gleichem Substitutionsgrad. Zur weiteren Klärung dieser Substituenteneinflüsse wurde geprüft, inwieweit sich eine chemische Modifizierung auf die Sorption von Cellulase an eine Celluloseoberfläche auswirkt. Die Untersuchungen trugen insofern Modellcharakter, als sie in einem heterogenen System durchgeführt wurden, bei dem neben der chemischen auch die physikalische Struktur der sorbierenden Oberfläche eine Rolle spielt. Sie sollten jedoch trotzdem gewisse indirekte Schlußfolgerungen auf die Enzymbindung im homogenen System ermöglichen.

Unter Verwendung der in [3] beschriebenen Methodik untersuchten wir die Sorption von *Penicillium*-Cellulase (Kulturfiltrat von *Penicillium citreoviride* HU Pc 2712 [4]) an verschiedene chemisch zu einem „Degree of substitution" (DS) von ca. 0,1 modifizierte, in Wasser quellbare, aber nicht lösliche Cellulosepulver.

Als Ausgangsmaterial diente ein hydrolytisch abgebautes und mechanisch nachzerkleinertes Cellulosepulver (Filtrak® „FNS" des VEB Spezialpapierfabrik Niederschlag/Erzgebirge), das — wie in den jeweiligen Literaturzitaten näher beschrieben — carboxymethyliert [5], sulfatisiert [6], cyanethyliert [7] bzw. zu Diethylaminoethylcellulose DEAE-Cellulose umgesetzt wurde. Zur Sorptionsuntersuchung wurden jeweils 0,5 g des Cellulosepräparates mit 10 ml Kulturfiltrat versetzt und nach einer Einwirkungszeit von 10 min/20 °C bzw. 30 min/0 °C abfiltriert und im Filtrat die „Restaktivität" der Enzymlösung ermittelt [3].

Die Aktivitätsbestimmung vor und nach der Sorption erfolgte im Hinblick auf „Linters abbauende Gesamtaktivität" [3], CMCL-Aktivität (Carboxymethylcellulose liquifying) [3], Cellobiaseaktivität [3], sowie den Gesamt-Proteingehalt der Enzymlösung [3]. Bei der Messung der „Linters abbauenden Aktivität" wurde außer dem in Lösung gebrachten Celluloseanteil auch der „Degree of Polymerisation" (DP) des Rückstandes ermittelt. Bei der Bestimmung der Cellobiase-Aktivität wurde die aus dem Sorbens freigesetzte Glucosemenge berücksichtigt. Die Heranziehung mehrerer unterschiedlicher Aktivitätskennwerte erwies sich für das hier vorliegende Multienzymsystem als zweckmäßig, um einerseits die Wirksamkeit des Gesamtkomplexes gegenüber in Faserform vorliegender Cellulose, andererseits die Aktivität einzelner Enzymkomponenten (Endoglucanase über CMCL-Aktivität; Cellobiase) zu charakterisieren.

Vorversuche mit unmodifiziertem „FNS“-Pulver ergaben nach einer Sorptionszeit von 10 min bei 20 °C keine signifikante weitere Aktivitätsabnahme bei Verlängerung der Sorptionsdauer. Die wesentlichen Ergebnisse unserer systematischen Untersuchungen sind in Tabelle 1 zusammengefaßt, in der jeweils die relative Höhe der einzelnen Aktivitätswerte nach dem Sorptionsvorgang, bezogen auf die Ausgangsaktivität = 100%, aufgeführt ist. Direkt vergleichbar auf Grund ihrer nur geringen Quellung in Wasser sind dabei die 3 Präparate „FNS unmodifiziert“, „CMC-Pulver“, „Cyanoethylcellulose-Pulver“. Die DEAE-Cellulose wies eine mittlere, das Cellulosesulfat eine sehr starke Quellung in Wasser auf. Durch den Sorptionsprozeß erfolgt im allgemeinen keine signifikante Verringerung der Cellobiase- und der CMCL- bzw. Endoglucanase-Aktivität, was auch in einigen orientierenden Messungen mit FNS-, CMC- und Cyanoethylcellulose-Pulver bei 0 °C und 30 min Sorptionszeit bestätigt wurde. Eine Ausnahme bildet lediglich die erhebliche Abnahme der Cellobiaseaktivität nach Kontakt mit dem stark

Tabelle 1. Sorption von Cellulase (*Penicillium citreoviride*) an modifizierten Cellulosepulvern. Sorptionstemperatur: 20 °C, Sorptionszeit: 10 Minuten

Sorbens	Proteingehalt (% vom Ausgangswert)	Cellobiase-aktivität (% vom Ausgangswert)	CMCL-Aktivität (% vom Ausgangswert)	Linters abbauende Aktivität	
				gelöster Anteil (% vom Ausgangswert)	DP-Rückstand
ohne	100	100	100	100	820
mikrokristallines Cellulosepulver	58	~ 100	98,8	83	1228
Carboxymethyl-cellulose	62	~ 100	99,5	100	1023
Cyanoethylcellulose	53	~ 100	97,9	71	1084
DEAE-Cellulose	57	~ 100	99,0	55	810
Cellulosesulfat	66	79	98,8	44	1073

quellenden Cellulosesulfat. Die Ursache hierfür sehen wir in erster Linie in der hohen Quellung bzw. großen zugänglichen Oberfläche dieses Präparates im Vergleich zu allen anderen, in Übereinstimmung mit einem früher [3] durchgeführten Vergleich des Sorptionsvermögens von unbehandelten und kolloid gemahlenen Baumwoll-Linters. Eine Abstufung im Rückgang der CMCL-Aktivität bei Einsatz unterschiedlich modifizierter Cellulosepulver als Sorbentien ist allenfalls als Tendenz erkennbar, indem diese Abnahme mit CMC-Pulver noch geringer ist als bei den anderen Sorbentien.

Im Gegensatz zu CMCL- und Cellobiase-Aktivität wird die „Linters abbauende Aktivität“ und der Proteingehalt der Enzymlösung durch Kontakt mit den verschiedenen Sorbentien in allen Fällen deutlich, aber in unterschiedlichem Maße erniedrigt. In Übereinstimmung mit unseren früheren Beobachtungen an unmodifizierten Cellulosepräparaten unterschiedlicher physikalischer Struktur [3] besteht dabei kein genereller Zusammenhang zwischen Abnahme des Proteingehaltes und Abnahme der „Linters abbauenden Aktivität“ der Enzymlösung, was erneut darauf hinweist, daß ein wesentlicher Anteil der im Kulturfiltrat vorhandenen Proteine keine cellulolytische Aktivität besitzen.

Die sorbierte Proteinmenge je Gramm eingesetzten Sorbens lag in unseren Versuchen zwischen 36,6 mg (Cellulosesulfat) und 50,6 mg (Cyanoethylcellulose), verglichen mit

Werten von PEITERSEN u. a. [10], die 19—88 mg Protein je Gramm Cellulose und SHAWKY u. a. [3], die 1,2—11,7 mg Protein je Gramm Cellulose sorbiert fanden. Die sorbierte Proteinmenge ist dabei abhängig vom eingesetzten Sorbens, der Sorptionszeit und Sorptionstemperatur sowie auch vom Flottenverhältnis Sorptionslösung/Sorbens [3, 10].
Die im Vergleich zum unmodifizierten FNS-Pulver geringere Abnahme der „Linters abbauenden Aktivität" bei Einsatz des CMC-Pulvers als Sorbens wurde durch ergänzende Sorptionsversuche bei 0 °C in der Tendenz bestätigt. Die sehr starke Aktivitätsabnahme mit DEAE-Cellulose könnte sowohl mit den kationischen Ladungen, als auch mit der stärkeren Quellung des Präparates zusammenhängen. Für die starke sorptionsbedingte Aktivitätsabnahme bei Einsatz von Cellulosesulfat sehen wir vor allem die große zugängliche Oberfläche (sehr starke Quellung) dieses Präparates als maßgebenden Faktor an. Zwischen Abnahme der „Linters abbauenden Aktivität" (ermittelt über die in Lösung gebrachte Substratmenge) und DP des Substratrückstandes zeigt sich in der Serie der hier untersuchten Sorbentien kein Zusammenhang. Offensichtlich hängt dieser DP-Wert sehr stark von dem Anteilsverhältnis unterschiedlicher Enzymkomponenten bzw. Inhibitoren der Cellulolyse ab, das beim Kontakt mit unterschiedlichen Sorbentien in unterschiedlicher Weise verändert wird.
Insgesamt führen unsere Untersuchungen zu dem Schluß, daß die sorptionsbedingte Abnahme von „Linters abbauender Aktivität" und Proteingehalt zwar durch funktionelle Gruppen des Sorbens beeinflußt wird, die physikalische Struktur des Sorbens jedoch eine mindestens ebenso große, wenn nicht größere Rolle spielt. Ein eindeutiger Zusammenhang zwischen Ladungszustand des Sorbens und Aktivitätsabnahme läßt sich aus unseren Ergebnissen nicht ableiten, wenn sich auch als Tendenz eine geringere Enzymsorption am CMC-Pulver im Vergleich zum unmodifizierten FNS-Pulver abzeichnet. Bemerkenswert erscheint die sehr geringe Sorption der die CMCL-Aktivität hervorrufenden Enzymkomponenten an sämtlichen hier untersuchten Cellulosepulvern, obwohl nach [8] bei der heterogenen enzymatischen Hydrolyse ein ausgeprägter Synergismus von Endo- und Exoglucanasen besteht und obwohl generell angenommen wird, daß unter Homogenbedingungen die enzymatische Spaltung der Celluloseketten durch diese mit der CMCL-Aktivität erfaßten Endoglucanasen erfolgt. Im Hinblick auf die homogene enzymatische Hydrolyse niedrig substituierter Cellulosederivate und den Einfluß von Art des Substituenten und DS auf den Hydrolyseverlauf könnte man hiernach vermuten, daß für den in [2] beschriebenen Einfluß der Art der funktionellen Gruppe auf den Hydrolyseverlauf nicht in erster Linie Richtung und Größe COULOMB'scher Wechselwirkungen, sondern die Möglichkeit der Ausbildung einer für das Entstehen des Enzym-Substrat-Komplexes günstigeren Kettenkonformation ist. Diese Überlegungen stimmen auch mit Vorstellungen zur Bildung des Enzym-Substrat-Komplexes überein [9], wonach die optimale Kettenlänge bei der Hydrolyse von Cellooligosacchariden mit einer gereinigten Endoglucanase 6 Glucoseeinheiten beträgt.

Eingegangen: 16. 7. 1984

Literatur

[1] Philipp, B., Kasulke, U., Lukanoff, B., Jacopian, V., Polter, E.: Acta Polymerica **33** (1982), 714.
[2] Philipp, B., Kasulke, U., Dautzenberg, H., Polter, E., Hübert, S.: Acta Polymerica **34** (1983), 651.
[3] Shawky, B., Kasulke, U., Philipp, B., Schulz, G., Hirte, W.: Acta Biotechnol. **4** (1984), 267.
[4] Hirte, W. F., Glatke, I.: WP C 12 d/189403.
[5] Dautzenberg, H., Philipp, B., Zöbisch, B.: Zellstoff und Papier **29** (1980), 156, 201.
[6] Philipp, B., Wagenknecht, W.: Cellul. Chem. Technol. **17** (1983), 443.
[7] Lukanoff, B., Dautzenberg, H., Philipp, B.: Acta Polymerica **30** (1979), 569.
[8] Eriksson, K.-E., Petterson, B.: Eur. J. Biochem. **51** (1975), 193, 312.
[9] Li, L. H., Flora, R. M., King, K. W.: Arch. Biochem. Biophys. **111** (1965), 439.
[10] Peitersen, N., Medeiros, J., Mandels, M.: Biotechnol. Bioeng. **19** (1977), 1091.

Acta Biotechnol. 5 (1985) 3, 313—317

Antiphytovirale Aktivität von lipophilen Fraktionen aus der Hefe Lodderomyces elongisporus IMET H 128

VOIGT, B[1], MÜLLER, H.[1] und SCHUSTER, G.[2]

[1] Akademie der Wissenschaften der DDR
Institut für Biotechnologie
DDR-7050 Leipzig, Permoserstraße 15
[2] Karl-Marx-Universität Leipzig
Sektion Biowissenschaften/Pflanzenphysiologie und Mikrobiologie
DDR-7010 Leipzig, Talstraße 33

Vortrag auf dem Leipziger Biotechnologiesymposium, 10.—14. September 1984

Summary

Lipid extract isolated from biomasses of *Lodderomyces elongisporus* IMET H 128 grown on gas oil (B.p. 240—380 °C) and fractions of the lipid extract (acetone soluble and fatty acid fractions) reduced the concentration of potato virus X in inoculated as well as in secondarily infected leaves markedly. The antiphytoviral inactive phosphatide fraction was converted into a fraction with high activity by partial hydrolysis with SO_2 as acting agent.

Die mikrobielle Gewinnung von Futterhefe auf der Basis von Kohlenwaserstoffen erfordert eine extraktive Reinigung der Biomassen von nicht utilisierten Kohlenwasserstoffen. Die ähnlichen Bindungsstrukturen und Löslichkeiten von Kohlenwasserstoffen und Lipiden führen dazu, daß die Extraktion mit der gleichzeitigen Entfernung von lipophilen zelleigenen Bestandteilen verbunden ist. Im Ergebnis der Extraktion werden deshalb Lipidextrakte erhalten, die sich aus substratspezifischen Kohlenwasserstoffen und stamm- und züchtungsbedingten Hefelipiden zusammensetzen [1].
Am Beispiel des Hefestammes *L. elongisporus* (Kohlenstoff-Quelle: Erdöldestillat der Siedelage 240—380 °C) wurde untersucht, ob derartige Lipidextrakte und daraus isolierte Fraktionen auf Grund ihrer strukturellen Spezifität als Präparate für die Bekämpfung pflanzlicher Virosen wirksam sind.
In Tab. 1 ist die stoffliche Zusammensetzung der aus *L. elongisporus* isolierten Lipidextrakte angegeben [2]. Durch einfache Trennoperationen läßt sich der Lipidextrakt in Fraktionen auftrennen (Abb. 1). Durch Aceton wird eine Auftrennung in acetonlöslichen und acetonunlöslichen Anteil erreicht. Das Acetonunlösliche wird durch weitere Raffi-

Tabelle 1. Zusammensetzung der aus der Hefe *Lodderomyces elongisporus* IMET H 128 isolierten Lipidextrakte

Komponenten	Konzentration in %
Phosphatide	25—35
Glyceride	15—25
Fettsäuren	5—10
Ergosterol	0,6—0,8
Ubichinon-9	0,1—0,2
Kohlenwasserstoffe	35—45

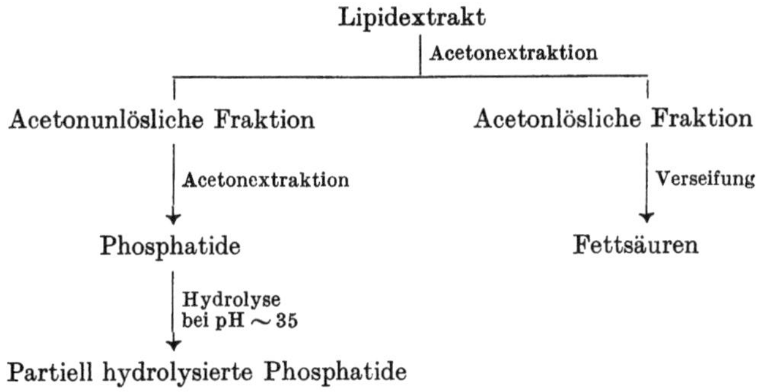

Abb. 1. Gewinnung von Lipidfraktionen

nierung mit Aceton zu Reinphosphatiden aufgearbeitet. Durch saure Hydrolyse — beispielsweise mit SO_2 — wird eine partiell hydrolysierte Phosphatidfraktion erhalten. Aus dem acetonlöslichen Anteil werden nach Verseifung die Fettsäuren isoliert.

Die Fettsäuren weisen ein substrat- und stammabhängiges Spektrum auf (Abb. 2), das sich über den Kettenlängenbereich C_{12} bis C_{18} erstreckt. Auffallend ist dabei der hohe Anteil einer ungesättigten C_{17}-Fettsäure. Mit ca. 45 bzw. ca. 65% ist der Anteil an ungeradzahligen bzw. ungesättigten Fettsäuren hoch.

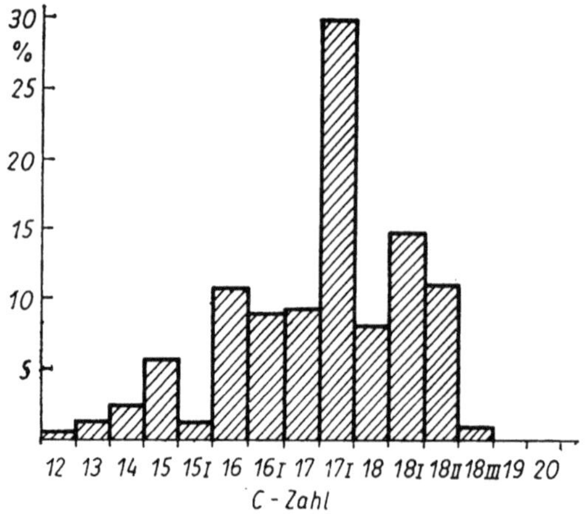

Abb. 2. Zusammensetzung der Fettsäurefraktion

Hauptbestandteile der Reinphosphatide sind Phosphatidylcholin und Phosphatidylethanolamin neben geringeren Anteilen ihrer Lysoverbindungen, Phosphatidylinosit, Phosphatidylserin und Cardiolipin (Tab. 2). Die in den Phosphatiden gebundenen Fettsäuren weisen ein mit den Gesamt-Fettsäuren vergleichbares Spektrum auf [3].

Lipidextrakt und daraus isolierte Fraktionen wurden in bekannter Weise auf antiphytovirale Aktivität gegenüber Kartoffel-X-Virus und Tabakmosaikvirus geprüft [4]. Verfolgt wurde die Wirksamkeit sowohl in primär als auch in sekundär infizierten Blättern von Nicotiana tabacum ‚Samsun'.

Lipidextrakt, Acetonlösliches und Fettsäuren vermindern die Konzentration des Kartoffel-X-Virus (Tab. 3). Je niedriger die relativen Virusgehalte sind, desto stärker ist die Hemmwirkung. Besonders stark ist die Hemmwirkung in sekundär infizierten Blättern.

Tabelle 2. Zusammensetzung der Phosphatidfraktion

Komponenten	Konzentration in %
Phosphatidylcholin einschließlich Lysophosphatidylethanolamin und Phosphatidylinosit	60,7
Phosphatidylethanolamin	27,7
Lysophosphatidylcholin Phosphatidylserin	5,9
Cardiolipin	4,7

Tabelle 3. Veränderung der Konzentration von Kartoffel-X-Virus nach Behandlung von Tabakpflanzen (Nicotiana tabacum ‚Samsun') mit Lipidfraktionen

Lipidfraktion	Konzentration der Präparation [%]	relativer Virusgehalt (Kontrolle = 100%)	
		inokulierte Blätter	sek. infizierte Blätter
Lipidextrakt	0,1	57+	30+
	0,5	77+	20+++
Acetonlösliche	0,1	64·	51+++
Fraktion	0,5	79·	34+++
Fettsäuren	0,1	69·	20++
	0,5	82·	20++

Lösungsmittel: 0.25 Vol.-% Tween 60 bzw. 5 Vol.-% Aceton

Signifikanzsymbole:

· : $p > 5\%$
+ : $5\% \geqq p > 1\%$
++ : $1\% \geqq p > 0{,}1\%$
+++ : $0{,}1\% \geqq p$

Tabelle 4. Hemmung der Vermehrung von Kartoffel-X-Virus in inokulierten Blättern von Nicotiana tabacum ‚Samsun' durch Kombinationen von Lipidfraktionen mit 2,4-Dioxohexahydro-1,3,5-triazin (DHT)

Lipidfraktion	relativer Virusgehalt (Kontrolle = 100%)		
	Lipidfraktion	DHT	Kombination
Lipidextrakt	57+ (0,1%)	61+ (0,0025%)	50++
Acetonlösliche Fraktion	82· (0,1%)	90· (0,005%)	67+
Fettsäuren	73· (0,1%)	58++ (0,0025%)	54++

Lösungsmittel: 0,25 Vol.-% Tween 60 bzw. 5 Vol.-% Aceton

Die Wirksamkeit der geprüften Fraktionen ist graduell unterschiedlich. Beispielsweise vermindert die Fettsäurefraktion die Konzentration des Kartoffel-X-Virus in sekundär infizierten Blättern bereits bei einer Aufwandmenge von 0,1% in etwa gleichem Umfang wie Lipidextrakt in einer Aufwandmenge von 0,5%. Kombinierte Ausbringung zusammen mit hydrierten Triazinen und Guanidinen steigert die antiphytoviralen Wirkungen. Therapeutisch bedeutsam ist die Wirkungssteigerung von Lipidextrakt in primär infizierten Blättern durch 2,4-Dioxohexahydro-1,3,5-triazin (Tab. 4). In sekundär infizierten Blättern wird die Viruskonzentration bei kombinierter Behandlung z. T. bis unter die serologische Nachweisgrenze bzw. bis dicht an diese abgesenkt (Tab. 5).

Tabelle 5. Hemmung der Vermehrung von Kartoffel-X-Virus in sekundär infizierten Blättern von Nicotiana tabacum ‚Samsun' durch Kombination von Lipidfraktionen mit 2,4-Dioxohexahydro-1,3,5-triazin (DHT)

Lipidfraktion	relativer Virusgehalt	(Kontrolle = 100%)	
	Lipidfraktion	DHT	Kombination
Lipidextrakt	30++ (0,1%)	5+++ (0,0025%)	0+++
Acetonlösliche Fraktion	51++ (0,1%)	4+++ (0,005%)	1+++
Fettsäuren	79+ (0,5%)	20+++ (0,0025%)	0+++

Lösungsmittel: 0,25 Vol.-% Tween 60 bzw. 5 Vol.-% Aceton

Phosphatide unterschiedlichen Reinheitsgrades und unterschiedlicher Zusammensetzung wiesen keine bzw. nur geringe antiphytovirale Wirksamkeit auf. Erst die partiell hydrolysierten Phosphatide verhinderten oder verlangsamten die Vermehrung von Kartoffelviren (Tab. 6). Kombinierte Ausbringung zusammen mit hydrierten Triazinen und Guanidinen steigert die antiphytovirale Wirkung beträchtlich und ist therapeutisch bedeutsam (Tab. 7). Darüber hinaus wird durch die Anwendung von partiell hydrolysierten Phosphatiden die Zahl der Lokalläsionen des Tabakmosaikvirus auf Nicotiana glutinosa beträchtlich vermindert.

Tabelle 6. Veränderung der Konzentration von Kartoffel-X-Virus in inokulierten und sekundär infizierten Blättern von Nicotiana tabacum ‚Samsun' nach Behandlung mit partiell hydrolysierten Phosphatiden

Pflanzenteil	Konzentration der Präparation [%]	relativer Virusgehalt (Kontrolle = 100%)
inokulierte Blätter	0,1[2]	80·
inokulierte Blätter	0,5[2]	41+++
sek. infizierte Blätter	0,1[1]	20++
sek. infizierte Blätter	0,5[1]	20++
sek. infizierte Blätter	0,1[2]	67·
sek. infizierte Blätter	0,5[2]	49++

Lösungsmittel: 5 Vol.-% Aceton[1] bzw. 0,25 Vol.-% Twenn 60[2]

Tabelle 7. Hemmung der Vermehrung von Kartoffel-X-Virus durch Kombination von partiell hydrolysierten Phosphatiden (PHP, 0,1%) mit 2,4-Dioxohexahydro-1,3,5-triazin (DHT, 0,0025%) bzw. Cyanoguanidin (CG, 0,1%)

Partner	Pflanzenteil	relativer Virusgehalt (Kontrolle = 100%)		
		PHP	Partner	Kombination
DHT	inok. Blätter	70++	83·	30+++
CG	inok. Blätter	70++	77++	40+++
DHT	sek. infizierte Blätter	88·	26+++	5+++
CG	sek. infizierte Blätter	88·	109·	76·

Lösungsmittel: 0,25 Vol.-% Tween 60

Der Umfang der nachgewiesenen antiphytoviralen Aktivitäten läßt erwarten, daß die getesteten lipophilen Fraktionen, die nicht toxisch und mikrobiell abbaubar sind, als neue antiphytovirale Chemotherapeutika geeignet sind und in der landwirtschaftlichen Praxis Anwendung finden werden. Gleichzeitig können Nebenprodukte biotechnologischer Verfahren einer effektiven volkswirtschaftlichen Nutzung zugeführt werden.

Eingegangen: 17. 10. 1984

Literatur

[1] Voigt, B. u. a.: Chem. Tech. **31** (1979), 409.
[2] Müller, H., Voigt, B.: Acta Biotechnol. **1** (1981), 279.
[3] Seidel, H. u. a.: Die Nahrung **27** (1983), 139.
[4] Schuster, G.: Archiv Phytopath. u. Pflanzenschutz **13** (1977), 231.

Acta Biotechnol. 5 (1985) 3, 318

Book Review

J. M. Franz; A. Krieg

Biologische Schädlingsbekämpfung

Schriftenreihe „Pareys Studientexte" Nr. 12

Berlin-West, Hamburg: Verlag Paul Parey. 1982.
252 S., 46 Abb., 8 Tab., 34 DM

Dieses Buch muß allen willkommen sein, die sich besorgt fragen, wie wohl der Mensch seine produktiven Stoffwandlungen, die die gesamte Biosphäre erfassen, fortsetzen, womöglich steigern und dennoch Mitglied natürlicher Lebensgemeinschaften bleiben kann. Es werden 17 knappe Sektionen in angewandter Ökologie geboten.

Untersucht wird die Frage, wie sich natürliche Biozönosen im Gleichgewicht halten, welche dieser Mechanismen in Produktionslandschaften, wie Feldern, Wäldern und Obstplantagen, noch wirksam sind und wie man sie nutzen kann. Die Beispiele reichen von der mäusejagenden Katze bis zum Einsatz von Sexualpheromonen. Verdeutlicht werden die Vorteile biologischer Schädlingsbekämpfung: hohe Spezifität, keine Resistenzen, keine Rückstände, keine Nebenwirkung. Wo Mikroben als natürliche Antagonisten, vor allem gegenüber Makroorganismen eingesetzt werden können, stellt sich die Frage ihrer Massenproduktion. Zunehmende Bedeutung gewinnen Signalstoffe, die ihre Wirkung als Attraktantien, Phagostimulantien, Repellentien, Pheromone, Wachstums- oder Entwicklungshormone ausüben. Ihre Erzeugung ist schon heute eine Aufgabe, deren sich die Gentechnik angenommen hat.

Ein Kapitel ist dem integrierten Pflanzen- und Gesundheitsschutz gewidmet. Das ist eine Verfahrensweise, bei der chemische und biologische Methoden kombiniert werden, um Schadorganismen unter der wirtschaftlichen Schadensschwelle zu halten.

Die Verfasser haben die sehr komplexen Zusammenhänge so klar und übersichtlich dargestellt, daß das Buch dem professionellen Schädlingsbekämpfer ebenso von Nutzen ist wie dem in Umweltfragen engagierten Laien.

Es ist eine Fundgrube überraschender und ermutigender Informationen und eine Anleitung zum Handeln für den Biologen, den Biotechnologen und den Landwirt.

K. Sattler

Acta Biotechnologica

Volume 5 1985 Number 3

Contents

Short Communications

Acta Biotechnologica is indexed in Current Contents/ET & AT; Chemical Abstracts; Biotechnology Abstracts; Excerpta Medica

Bei Fragen zur Produktsicherheit wenden Sie sich bitte an:
If you have any questions regarding product safety,
please contact:

Walter de Gruyter GmbH
Genthiner Straße 13
10785 Berlin
productsafety@degruyterbrill.com